Slicing Pizzas,
Racing Turtles,
and Further Adventures in
Applied Mathematics

ROBERT B. BANKS

Slicing Pizzas, Racing Turtles,

and Further Adventures in Applied Mathematics

Princeton University Press
Princeton, New Jersey

Copyright © 1999 by Princeton University Press
Published by Princeton University Press, 41 William Street,
Princeton, New Jersey 08540
In the United Kingdom: Princeton University Press,
Chichester, West Sussex

Library of Congress Cataloging-in-Publication Data
Banks, Robert.
Slicing pizzas, racing turtles, and further adventures in
applied mathematics / Robert B. Banks.
p.   cm.
Includes bibliographical references and index.
ISBN 0-691-05947-0 (cloth : alk. paper)
1. Mathematics—Popular works. I. Title.
QA93.B358 1999
510—dc21   98-53513

This book has been composed in Times Roman

The paper used in this publication meets
the minimum requirements of
ANSI/NISO Z39.48-1992 (R1997)
(Permanence of Paper)

http://pup.princeton.edu

Printed in the United States of America

10 9 8 7 6 5 4 3 2 1

To my mother,
Georgia Corley Banks,

and my sister and brothers,
Joan, Dick, and Barney

Contents

Preface

In large measure, this book is a sequel to an earlier volume entitled *Towing Icebergs, Falling Dominoes, and Other Adventures in Applied Mathematics.* As in the previous work, this book is a collection of topics characterized by two main features: the topics are fairly easy to analyze using relatively simple mathematics and for the most part, they deal with phenomena, events, and things that we either run across in our everyday lives or can comprehend or visualize without much trouble.

Here are examples of a few of the topics we consider:

You need to go from here to there in a pouring rainstorm. To get least wet, should you walk slowly through the rain or run as fast as you possible can?

What blast-off velocity do you and your spacecraft need to entirely escape the earth's gravitational pull?

The colors of America's flag are, of course, red, white, and blue. Which of the three colors occupies the largest area of the flag and which color the smallest?

What is the length of the seam on a baseball or the groove on a tennis ball?

What is the surface area of the Washington Monument and what is its volume? How many golf balls could you put into an entirely empty Washington Monument?

Some of these questions sound trivial, perhaps even silly. Even so, they do depict settings or situations that are easy to visualize and understand. With the help of mathematics, it is not difficult to obtain the answers. As we shall see, the level of mathematics ranges from algebra and geometry to calculus. Several problems involve spherical trigonometry.

Throughout the book, topics involving various fields of knowledge are investigated. For example, quite a few problems featuring geography and demography are examined. In other chapters, a number of topics concerned with hydrology, geomorphology, and cartography are analyzed. In addition, where it is appropriate and feasible, features are described that relate to the historical aspects of a particular topic.

For over four decades, I was engaged in teaching and research at several universities and institutes in the United States and abroad (England, Mexico, Thailand, the Netherlands). I collected most of the topics presented in the book during that period. My primary interests, as a professor of engineering, were in the fields of fluid mechanics and solid mechanics (statics, dynamics, mechanics of materials).

Although this book deals with mathematics, it is certainly not intended to be a textbook. It might, however, be a worthwhile supplement to a text at the high school and university undergraduate levels.

As was the case in my earlier book, my strong hope is that this collection of mathematical stories will be interesting and helpful to people who long ago completed their formal studies. I truly believe that many of these "postgraduate students" sincerely want to strengthen their levels of understanding of mathematics. I think that this is especially true as all of us enter a new century that assuredly will place heavy emphasis on mathematics, science, and technology.

Here are some more examples of topics we examine in the book.

A prime number is a number that can be divided only by one and by itself. Some examples are 1, 2, 3, 5, 7, 11, 13, 17, and so on. As you might want to confirm, there are 168 prime numbers less than 1,000. Can you guess the magnitude of the largest prime number known at the present time (1999)?

A beautiful Nautilus sea shell has a shape called a logarithmic spiral. This attractive curve is mathematically related to the well-known Fibonacci sequence and the ubiquitous golden number, $\phi = (1/2)(1 + \sqrt{5})$.

The most famous "numbers" in all of mathematics are π (the ratio of the circumference of a circle and its diameter) and e (the base of natural logarithms). It is interesting that these two important numbers are related by the equation $e^{i\pi} = -1$ where $i = \sqrt{-1}$. The numerical values of π and e (to five decimal places) are $\pi = 3.14159$ and $e = 2.71828$. Would you believe that at the present time (1999), π has been calculated to more than 51 billion decimal places?

The number of people in the world is approximately 6.0 billion as we begin the new century. Is this a large percentage or a small percentage of the number of people who have ever lived on earth?

Most of the ice in the world is in Antarctica and Greenland. If all this ice melted, the oceans would rise by 75 meters (246 feet). What would this increase in water level do to Florida, Washington, D.C., the Mississippi River, and Niagara Falls?

Finally, I hope you will enjoy going through the book. In attempts to make things a bit easier, I have tried to be somewhat light-hearted here and there. We all know that mathematics is a serious subject. However, this does not mean we cannot be a little frivolous now and then.

Acknowledgments

Numerous people gave me considerable assistance during the period of preparation of this book. I am grateful to the following persons for their willingness to read various chapters of my manuscript: Philip J. Davis, J. Donald Fernie, Paulo Ribenboim, O. J. Sikes, Whitney Smith, and John P. Snyder.

Others aided in a substantial way by providing information dealing with all kinds of things. For this help, I would like to express appreciation to Teresita Barsana, Brian Bilbray, Roman Dannug, Bill Dillon, and Susan Harris.

As before, I express much gratitude to my editor, Trevor Lipscombe, and to many others at Princeton University Press, for their greatly appreciated efforts to turn my manuscript into a book.

It is impossible to list the many contributions my wife, Gunta, made to this endeavor. I thank her for all she has done. In my earlier book, her wise advice and cheerful help certainly improved the likelihood of a successful towing of that literary iceberg. This time, once again, her wise counsel and cheerful assistance surely strengthen the chances of a creditable race by her husband, a literary turtle.

Slicing Pizzas,
Racing Turtles,
and Further Adventures in
Applied Mathematics

1

Broad Stripes and Bright Stars

These days we see much more of the flag of the United States than we ever did in the past. Old Glory flies over many more office buildings and business establishments than it did before. It is now seen far more extensively in parks and along streets and indeed in a great many programs and commercials on television.

With this greatly increased presence and awareness of the flag, there is understandable growing interest in learning more about numerous aspects of the U.S. flag, including its history and physical features and the customs and protocol associated with it.

There are a number of books that examine various topics concerning the flag of the United States. Representative are the publications of Smith (1975a) and Furlong and McCandless (1981). On the world scene, many references are available dealing with the flags of all nations. For example, the books by Smith (1975b) and Crampton (1990) cover many subjects relating to flags of the world and to various other topics of *vexillology*: the art and science of flag study.

The Geometry of the Flag of the United States

With that brief introduction, we come directly to the point. The colors of the U.S. flag are red, white, and blue. Now, are you ready for the big question? What are the area *percentages* of red,

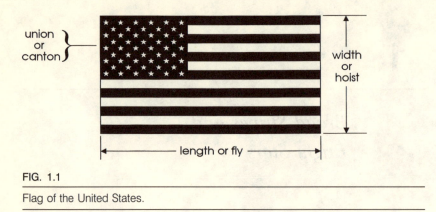

FIG. 1.1

Flag of the United States.

white, and blue? That is, which of the three colors occupies the largest area of the flag and which color the smallest? It's a good question. Do you want to guess before we compute the answer?

The flag is shown in figure 1.1 and its more important proportions and features are listed in table 1.1. Arbitrarily selecting the *foot* as the unit of linear measurement, here are some preliminary observations:

The total area of the flag is $1.0 \times 1.9 = 1.9$ ft^2

The area of the union is $7/13 \times 0.76 = 0.4092$ ft^2

The length of the seven upper stripes is 1.14 ft

The length of the six lower stripes is 1.9 ft

TABLE 1.1

Proportions and features of the U.S. flag

Item	Quantity
Width of flag	1.0
Length of flag	1.9
Number of red stripes	7
Number of white stripes	6
Width of union	7/13
Length of union	0.76
Number of stars	50
Radius of a star	0.0308

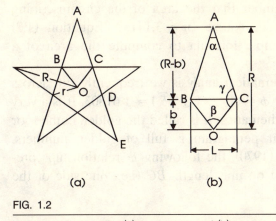

FIG. 1.2

Definition sketches for (a) pentagram and (b) pentagram kite.

The width of a stripe is $1/13 = 0.07692$ ft

The problem of computing the *red* area is easy

The problem of computing the *white* and *blue* areas is not so easy because of the 50 white stars in an otherwise blue union

So, before we can obtain the final answer we have to look at stars

The Geometry of a Five-Pointed Star

A five-pointed star, commonly called a *pentagram*, is shown in figure 1.2(a). Its radius, R, is the radius of the circumscribing circle. The five-sided polygon within the star is called a *pentagon*; the radius of its circumscribing circle is r.

The section *ABOC* is removed from the pentagram of figure 1.2(a) and displayed as the pentagram kite of figure 1.2(b). Some geometry establishes that $\alpha = 36°$, $\beta = 72°$, and $\gamma = 126°$. Without much difficulty we obtain the expression

$$A = \left[5 \frac{\sin(\alpha/2)\sin(\beta/2)}{\sin(\alpha + \beta)/2} \right] R^2, \qquad (1.1)$$

where A is the area of the pentagram. Substituting the values of α and β into this expression gives

$$A = 1.12257 R^2. \qquad (1.2)$$

For comparison, remember that the area of the circumscribing circle is πR^2 where, of course, $\pi = 3.14159$. Equation (1.2) provides us with a simple formula to compute the area of a five-pointed star.

In elementary mathematical analysis we frequently run across the numerical quantity $\phi = (1/2)(1 + \sqrt{5}) = 1.61803$. It is a very famous number in mathematics. It is called the golden number or divine proportion. Our pentagram is full of golden numbers. According to Huntley (1970), the following ϕ relationships prevail in figure 1.2 based on unit length BC (i.e., one side of the pentagon):

$$AE = \phi^3; \quad AD = \phi^2; \quad AC = \phi;$$

$$\frac{R}{r} = \phi^2; \quad \frac{R}{b} = 2\phi; \quad \frac{b}{r} = \frac{\phi}{2}; \quad \frac{L}{r} = \sqrt{3 - \phi}. \tag{1.3}$$

Utilizing these relationships, it can be established that the area of the regular pentagram can be expressed in terms of ϕ:

$$A = \left[\frac{5}{\phi^2} \sin(\beta/2) \right] R^2. \tag{1.4}$$

As we would expect from observing equation (1.2), the quantity in the brackets of equation (1.4) is 1.12257.

The perimeter of a pentagram is not difficult to determine. Using some geometry and trigonometry we obtain

$$P = \left[10 \frac{\sin(\beta/2)}{\sin(\alpha + \beta)/2} \right] R. \tag{1.5}$$

The bracketed quantity has the numerical value 7.2654. The length of the circumscribing circle is, of course, $2\pi R$.

Numerous other relationships could be established. For example, can you demonstrate that the ratio of the area of the five points of the pentagram to the area of the base pentagon is $\sqrt{5}$?

How Much Red, How Much White, How Much Blue?

We now have the information we need to answer the big question. From table 1.1 we note that the radius of a star is $R = 0.0308$ ft and so, from equation (1.2), the area of a single star is $A = 0.0010649$ ft^2. The area of 50 stars is 0.05325 ft^2.

Color: red

The red area, A_r, is

$$A_r = \frac{4}{13} \times (1.9 - 0.76) + \frac{3}{13} \times 1.9 = 0.78923 \text{ ft}^2.$$

Color: white

The white area, A_w, is

$$A_w = \frac{3}{13} \times (1.9 - 0.76) + \frac{3}{13} \times 1.9 + 0.05325$$

$$= 0.75479 \text{ ft}^2.$$

Color: blue

The blue area, A_b, is

$$A_b = (0.76 \times 7/13) - 0.05325 = 0.35598 \text{ ft}^2.$$

The total area of the flag is $A_{\text{flag}} = 1.9$ ft^2. So it is easy to calculate that the color percentages are 41.54% red, 39.73% white, and 18.73% blue. This means that if you are going to paint a really big U.S. flag, you will need 42 gallons of red paint, 40 gallons of white, and 19 gallons of blue to come out about even.

Here are several other items of information about the flag that you might want to confirm:

Based on a flag width of 1.0 foot, there are 10.26 feet of red stripe and 9.12 feet of white stripe. The total is 19.38 feet.

The union is 13.01% white and 86.99% blue.

If the 50 stars were replaced by a single big star with the same total area, it would have a radius of 0.218 feet.

The perimeter of a single star is 0.2238 feet; the total perimeter of all 50 is 11.19 feet. This is nearly double the perimeter of the entire flag.

This is sufficient. We stop here because we now have the information we set out to determine. That is, the red area is 41.6%, the white 39.7%, and the blue 18.7%.

Main Dimensions of Flags

The official flag of the United States has a ratio of width to length of 10:19. Why the relative length of the flag is precisely 1.9—or indeed why the relative length of the union is exactly 0.76 (could it be 1776?)—is not known. It just is. If you are interested in the historical aspects of vexillology, you might want to contact the Flag Research Center in Winchester, Massachusetts or the Flag Institute in Chester, England.

In any event, an interesting question has been raised by Nicolls (1987): "What are the ideal proportions of a flag?" Again, this is the kind of question for which there is no "scientific" answer. A brief history of the changes of flag dimension proportions is given by Nicolls. During the Middle Ages, for example, flag proportions ranged from an extremely short 1(width):0.5(length) ratio to a square 1:1 ratio.

Over the years, the poor visibility and inadequate flapping characteristics of quite short flags resulted in their gradual lengthening. At present, the official flags of the United Kingdom and thirty other nations have proportions of 1:2 and those of twenty five other countries possess ratios of 3:5. It is pointed out by Nicolls that these ratios are numerically close to that given by the golden mean, $1:\phi$, where $\phi = (1/2)(1 + \sqrt{5}) = 1.618$. We see that collectively the geometric proportions of the world's flags are not greatly different from the "divine proportion."

TABLE 1.2

Suggested countries for flag analysis

0 to 3 points	4 to 6 points	7 to 9 points	10 to 12 points
Bangladesh	Chile	Algeria	Guyana
Japan	Iceland	Cuba	Korea
Russia	Senegal	Namibia	Malaysia
Thailand	United Kingdom	Singapore	Nepal
Ukraine	Venezuela	Tunisia	New Zealand

A word about flapping or fluttering of flags. This is an interesting and not very easy problem in fluid mechanics. If you are challenged by the mathematics of the phenomenon you can study the section on "surface waves" in Lamb (1945). If you are interested in some experimental work, get yourself an electric fan, a few sheets of paper, a pair of scissors, and go to work.

PROBLEM 1. Now that you know all about the geometrical features of the flag of the United States, your assignment is to select and analyze the flag of some other country. As in the case of the U.S. flag, you are required to determine the percentage distribution of the colors of the flag you choose. A list of flags appears in table 1.2. Since some flags are geometrically simple and others are rather complicated, there is considerable spread in the number of points you will be awarded.

PROBLEM 2. If the width to length ratio of the U.S. flag were altered from 1:1.9 to 1:ϕ, and it was otherwise unchanged, what would be the area percentages of red, white, and blue?
Answer. Red: 39.4%, white: 38.6%, blue: 22.0%.

Revisiting the Stars

The pentagram—the five-pointed star—has a very long history. Its geometrical properties were known to the ancient Babylonians and it was regarded as a symbol and badge of the Grecian school of geometers in the days of Pythagoras.

We have seen the many relationships between the pentagram and the golden number, ϕ. Furthermore, as we shall learn later on, ϕ is the ratio of successive Fibonacci numbers $(1, 1, 2, 3, 5, 8, 13, \ldots)$ when these numbers become very large. So we note the connection between pentagrams and this famous mathematical sequence.

There are many interesting and practical problems involving pentagrams and pentagons; some are easy and some are difficult. For example, the area of the pentagon portion of a pentagram is

$$A = \left[\frac{5}{4}\sqrt{2 + \phi} \right] r^2 = 2.37762r^2, \tag{1.6}$$

where r is the radius of the circumscribing circle of the pentagon. In addition, the perimeter of a regular pentagon is

$$P = 5\sqrt{3 - \phi}\, r = 5.8779r. \tag{1.7}$$

For example, the famous Pentagon building near Washington has a side length $L = 921$ ft. Hence, the total perimeter is 4,605 feet. From equation (1.7) the radius $r = 783$ ft and so, from equation (1.6), the total area covered by the Pentagon is $A = 1,457,680$ ft^2. This is approximately 33.5 acres. Some questions: If they ever add the five points to the Pentagon building (thereby making it the Pentagram building) what will be the radius R, the perimeter P, and the total area A? The answers are that $R = 2,050$ ft, $P = 14,895$ ft, and $A = 108.3$ acres.

The Ubiquity of ϕ: Three Problems

In the preceding sections we have seen that the golden ratio, $\phi = (1/2)(1 + \sqrt{2}) = 1.61803$, makes numerous appearances in the patterns of pentagrams and pentagons. In later chapters, we take closer looks at this very interesting number. For the present, here are three problems that illustrate that ϕ shows up not only in the geometry of five-pointed stars but in many other places as well.

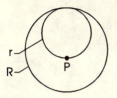

FIG. 1.3

Definition sketch for the golden earring problem.

GOLDEN EARRINGS.   A circle of radius r is removed from a larger circle of radius R, as shown in figure 1.3. The center of gravity of the remaining area is at the edge of the removed circle, P. Confirm the result obtained by Glaister (1996) that this "golden earring" indeed balances at point P, if the ratio of the radii of the two circles is $R/r = \phi$.

PENTAGON CIRCLES.   An interesting design or pattern for a flag might be the array of circles shown in figure 1.4. In this figure, five circles, each of radius r, are placed with their centers at the corners of a regular pentagon and their circumferences passing through the centroid of the pentagon, O.

   Show that the radius R of the largest circle that can be covered by the five smaller circles is given by $R/r = \phi$.

   This problem was originally posed by the noted English mathematician E. H. Neville in 1915; it is discussed by Huntley (1970).

FIG. 1.4

The problem of the pentagon circles.

FOLDED PENTAGRAM.   In figure 1.2, triangle ABC is folded about BC and the other four triangles are folded in the same fashion. The five sloping triangular sides meet at a point to create a "folded pentagram" pyramid with a pentagon base of radius r.

Show that the height of the pyramid is $H/r = \phi$ and its volume is

$$V = \left[\frac{5}{12} \phi \sqrt{2 + \phi} \right] r^3. \tag{1.8}$$

2

More Stars, Honeycombs, and Snowflakes

Skinny Pentagrams and Fat Pentagrams

So far, in our study of pentagrams—that is, five-pointed stars—we have considered only the quite familiar pentagram that appears on the flag of the United States and in so many other places. We shall say that the five points that appear on this particular pentagram are normal or "regular" in shape.

In contrast, our next step is to examine the geometry of pentagrams with points that range in shape from "long and skinny" to "short and fat." Let figure 2.1 serve as our definition sketch.

The main geometrical variable in our problem is the ratio r/R, in which R is the radius of the *external* circumscribing circle and r is the radius of the *internal* circumscribing circle. It is clear that the angle $\beta = 72°$ regardless of the value of r/R (because $5 \times 72° = 360°$). For small values of r/R, we have skinny pentagrams and for large values we have fat ones. In between, if $r/R = 1/\phi^2$, where $\phi = (1/2)(1 + \sqrt{5}) = 1.61803$, we have our familiar regular pentagram. These three cases are illustrated in figure 2.2.

It is logical to identify the regular pentagram as that pentagram for which $\alpha = 36°$ or, equivalently, $r/R = 1/\phi^2$. An array of adjectives is available to define the cases for which $\alpha < 36°$

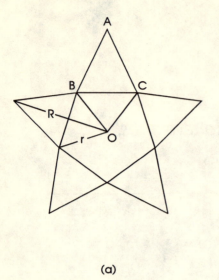

(a)

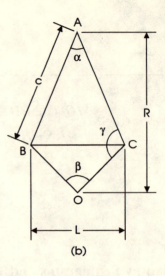

(b)

FIG. 2.1

Definition sketches for (a) pentagram and (b) pentagram kite.

(a)   (b)   (c)

FIG. 2.2

Pentagram shapes for (a) $r/R = 0.20$, (b) $r/R = 1/\phi^2 = 0.382$, and (c) $r/R = 0.70$.

(long and skinny) and $\alpha > 36°$ (short and fat). The terminology acute–regular–obtuse seems to be the simplest and most descriptive way to define the three categories of pentagrams.

Some Geometrical Features of Pentagrams

Utilizing figure 2.1 and the law of sines, we obtain the equations

$$\frac{r}{R} = \frac{\sin(\alpha/2)}{\sin\gamma}; \quad \frac{r}{c} = \frac{\sin(\alpha/2)}{\sin(\beta/2)}. \tag{2.1}$$

We want to express the various quantities in our pentagram problem in terms of the variable r/R. Letting $p = r/R$ and employing the above equations, we determine that the area of the pentagram is

$$A/R^2 = 5p\sin(\beta/2), \tag{2.2}$$

and the perimeter is

$$P/R = 10p\frac{\sin(\beta/2)}{\sin(\alpha/2)}. \tag{2.3}$$

The results of computations based on these equations are listed in table 2.1. We make the following observations:

Clearly, if $p = 1/\phi^2 = 0.38197$, the pentagram becomes the familiar five-pointed star.

When $p = 0.5$, the pentagram takes on the shape of the star used in the Berghaus projection of the world.

When $p = 1/\phi = 0.61803$, both α and β are 72° and the pentagram is composed of ten identical triangles.

When $p = \phi/2 = 0.80902$, the pentagram becomes a regular pentagon. As we observe in the table, the perimeter is a minimum for this value of p.

When $p = 1$, the pentagram becomes a regular decagon (i.e., a ten-pointed polygon).

TABLE 2.1

Geometrical features of pentagrams

$p = r / R$	α	γ	A / R^2	P / R
0	0	144	0	10.000
0.20	15.97	136.02	0.5878	8.464
0.382	36	126	1.1226	7.265
0.50	52.53	117.74	1.4695	6.641
0.618	72	108	1.8163	6.180
0.70	86.99	100.51	2.0573	5.978
0.809	108	90	2.3777	5.878
0.90	125.59	81.21	2.6450	5.948
1	144	72	2.9389	6.180

Note: $\alpha = 36°$ is the regular pentagram. $\beta = 72°$ for all values of r/R.

Berghaus Star Projection of the World

When $r/R = 0.5$, the radius of the internal polygon is one-half the radius of the pentagram. As mentioned earlier, this is the radius ratio of the Berghaus star projection of the world.

Later on, in chapter 22 on cartography, this projection is examined again. For now, we simply make the observation that in the Berghaus projection, the northern hemisphere of the world is displayed in the internal circumscribing circle of the star. The southern hemisphere appears in the five points of the star defined at 72° intervals along the equator. A recommended reference on this topic is Snyder (1993).

Six-Pointed Stars: Hexagrams and Hexagons

We conclude our study of the geometry of stars with a brief analysis of six-pointed stars or *hexagrams*. A definition sketch appears in figure 2.3. The radius of the hexagram is R. The six-sided polygon inside the hexagram is called a *hexagon*; its radius is r.

In comparison with the geometry of a five-pointed star—the pentagram—that of the six-pointed star is simple. This is because the hexagram and its associated hexagon are composed of noth-

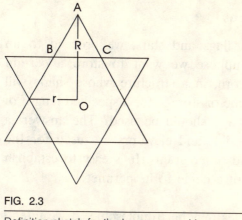

FIG. 2.3

Definition sketch for the hexagram and hexagon.

ing but equilateral triangles. In figure 2.3, *ABC* is a representative triangle. Since the length of each of its three sides is r and each of the internal angles is 60°, it is not difficult to establish that the area of each triangle is $(\sqrt{3}/4)r^2$ and the perimeter of each is $3r$.

With this information, we easily determine that the area and perimeter of a hexagon are

$$A = \left(3\sqrt{3}/2\right)r^2; \; P = 6r. \tag{2.4}$$

Likewise, the area and perimeter of a hexagram are

$$A = (3\sqrt{3})r^2; \; P = 12r. \tag{2.5}$$

Finally, we note that $R = \sqrt{3}\,r$ and hence the radius ratio is $p = r/R = 1/\sqrt{3} = 0.577$. The area and perimeter of the hexagram can be expressed in terms of radius R instead of r. The answers are

$$A = \sqrt{3}\,R^2; \; P = 4\sqrt{3}\,R. \tag{2.6}$$

PROBLEM   In the previous chapter, we determined the area percentages by color of the flag of the United States. It is observed that the Star of David, which appears on the flag of Israel, is a regular hexagram. What are the area percentages by color of this flag?

Honeycombs and Hexagons

Leaving the subjects of flags and stars, we proceed to an entirely different topic. Suppose we want to store a certain commodity, for example, corn, in a structure whose shape will provide the maximum volume of storage for a specified length of storage boundary. What shape should be used? The answer: a circle. That's why circular silos and elevators are used for the storage of corn, wheat, and other grains. It is easy to establish that the ratio of the area of a circle to its perimeter is $A/P = \pi r^2/2\pi r = r/2 = 0.500r$.

However, the use of circles may be troublesome. If we intend to build quite a few circular storage structures and place them side by side, we will have empty zones between the circles, no matter how we arrange them. In other words, circles are not "space-filling" shapes.

Well, what shapes are space-filling? The answer: equilateral triangles, squares, and hexagons. It is easy to establish that pentagons, octagons, and other n-gons do not qualify; they don't fit.

It is interesting to note that ordinary honeybees evidently know all about the geometry of optimized storage space. For a very long time indeed, these lively little insects have been constructing honeycombs for the storage of their very well-known commodity: honey. And what shape of structures do they utilize? If you said *hexagons*, you are absolutely correct.

Here is a historical note on this topic. Many years ago, the German astronomer-mathematician, Johannes Kepler (1571–1630), wrote an interesting little book entitled *The Six-Cornered Snowflake* as a New Year's gift for the emperor who provided his financial support. In his essay, Kepler (1611) raised the question of why snowflakes always have six sides. To quote him directly, "There must be a cause why snow has the shape of a six-cornered starlet. It cannot be chance. Why always six?"

As part of his wonderment concerning snowflakes, Kepler was curious about the six-cornered plan on which honeycombs are built. Again, to quote him:

> ...what purpose had God in putting these canons of architecture into the bees? Three possibilities can be imagined. The hexagon is the roomiest of the three plane-filling figures (triangle, square, hexagon); the hexagon best suits the tender bodies of the bees; also labour is saved in making walls which are shared by two; labour would be wasted in making circular cells with gaps.

Kepler concluded his short volume with a plea to chemists to study the problem and provide some answers. In addition, he presented a challenge "... to those who followed him to discover the mathematics of the emergence of visible forms in crystals, plants and animals."

It is noteworthy that currently mathematicians and scientists are devoting much attention to research on the topics advocated by Kepler. In this regard, an intriguing book by Neill (1993), entitled *By Nature's Design*, presents a collection of remarkable photographs illustrating patterns, form, and shape of things ranging from honeycombs and snowflakes to seashells and spider webs.

PROBLEM It was established in a preceding paragraph that the ratio of the area of a circle to its perimeter is $A/P = 0.500r$. Determine the value of this ratio for a hexagon, a square, and an equilateral triangle. On the basis of these results, list some reasons why you think honeybees wisely use hexagons as the basic shape for their honeycombs.

Snowflakes and Hexagons

The same questions asked by Kepler regarding the hexagonal shape of snowflakes have been raised by countless others over the years. Scientists working in the field of *crystallography* now understand the molecular structure of snowflakes and ice crystals and the reasons for their two-dimensional hexagonal shape. A suggested reference on this topic is Knight and Knight (1973). The interesting book by Bentley and Humphreys (1962) displays more than two thousand photographs of beautiful hexagonal snow crystals.

Another topic relating to hexagons and other polygons is the remarkable work of the Dutch artist, M. C. Escher (1898–1972). His spectacular paintings and drawings invariably feature intriguing and beautiful displays of symmetry.

In 1985, an international conference was held in Rome, attended by many scientists, mathematicians, artists, and historians, to commemorate Escher's work. The proceedings of the conference were published the following year with H.S.M. Coxeter serving as senior editor. Many of Escher's spectacular drawings appear in these proceedings, Coxeter et al. (1986). It also includes numerous articles about Escher and his work prepared by specialists in various fields of science, mathematics, and the humanities.

Although Escher had essentially no training in mathematics, he possessed an incredible understanding of geometrical principles and space perception. Some of his works, for example, are remarkably beautiful displays of triangular and hexagonal symmetry.

Recently, Coxeter, one of the leading geometers of the twentieth century, carried out a mathematical analysis of one of Escher's woodcuts. His paper, Coxeter (1996), is interesting reading; only elementary geometry and trigonometry are needed to understand it.

The Koch Snowflake and a Brief Introduction to Fractals

We conclude our chapter with a look at the so-called Koch snowflake. It has nothing to do with a real snowflake. It is called one simply because successive mathematical embellishments of an equilateral triangle create a shape that ends up looking like a snowflake. By the way, Helge von Koch was a Swedish mathematician. In 1904, he devised the problem we now consider.

The problem begins with the equilateral triangle shown in figure 2.4(a); the side length of the triangle is r. We shall call this

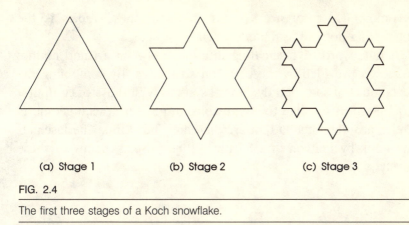

(a) Stage 1   (b) Stage 2   (c) Stage 3

FIG. 2.4

The first three stages of a Koch snowflake.

stage 1. In figure 2.4(b), stage 2, we attach an equilateral triangle of side length $r/3$, at the midpoint of each of the sides of stage 1. In our next step, shown in figure 2.4(c), stage 3, triangles are again attached at the midpoints, this time of side length $r/9$. The process is continued in this fashion for as long as we please. Two comments. First, noting that the shape gets fancier at each successive stage, we can understand why they call it a snowflake. Second, if we take a magnified look at the edge of our prickly snowflake, after a number of stages, we observe that there is no basic change in the geometric pattern of the boundary. Consequently, we say that the Koch snowflake is a *self-similar* curve.

PROBLEM With reference to figure 2.4, the area and perimeter of stage 1 are $A = (\sqrt{3}/4)r^2$ and $P = 3r$. Likewise, the area and perimeter of stage 2 are $A = (3\sqrt{3})(r^2/9)$ and $P = 4r$. Continuing through an infinite number of stages, show that the area of the snowflake is finite — and equal to $8/5$ of the area of the original triangle — but that its perimeter is infinite.

These remarks serve as a brief introduction to an interesting subject: *fractals*. Some aspects of this subject are certainly not new. For example, the snowflake devised by von Koch in 1904 attracted little attention until the classic publication of Mandelbrot (1982). Since then, scores of books have been published on the subject of fractals and the related topic of *chaos theory*. A recommended reference for the study of these topics is the interesting book by Çambel (1993).

A final word: A fascinating interface between art and mathematics is provided by these fractal structures. By combining the techniques of so-called dynamical systems with those of computer graphics, it is possible to create color displays of fractal geometry and chaos phenomena that are incredibly beautiful. The fascinating book by Peitgen and Richter (1986) displays many of these lovely pictures.

3

Slicing Things Like Pizzas and Watermelons

We Start with Pizza

Our problem begins with the supposition that you have a large pizza in front of you and you want to obtain the maximum number of pieces with a certain number of straight line slices. With one slice you get two pieces of pizza, two slices give you four pieces, and three slices get you six, right?

Not necessarily. If your third slice avoids the intersection of the first two slices, you will have a total of seven pieces.

A short pause while you get a pad of paper, a pencil, and a ruler. All set? Draw a straight line and then another that intersects the first. These are slices one and two and you have four pieces. Another line—slice three—gives seven pieces and slice four yields eleven. Slices five and six give sixteen and twenty-two, respectively.

By now it is getting to be a bit difficult to count the pieces. Never mind. We shall use what we now know. Let C be the number of slices or cuts and let P be the total number of pieces. Also let ΔP be the *difference* between successive values of P and let $\Delta^2 P$ be the *difference of the difference*. We compute these quantities, ΔP and $\Delta^2 P$, to determine whether a clear mathematical pattern develops as the P sequence increases. Thus, we have the numbers shown in table 3.1.

TABLE 3.1

The two-dimensional case ($n = 2$)

C	P	ΔP	$\Delta^2 P$
0	1		
		1	
1	2		1
		2	
2	4		1
		3	
3	7		1
		4	
4	11		1
		5	
5	16		1
		6	
6	22		

This is very interesting, especially the ΔP and the $\Delta^2 P$ columns. At this point, you probably discover what seems to be the rule to determine values of P for whatever value of C. The sequence of $\Delta^2 P$ is just a string of ones and the sequence of ΔP is simply ordinary counting. So, for $C = 7$ you predict $P = 29$ and for $C = 8$ you anticipate $P = 37$. You have indeed determined the sequence of P (viz., $1, 2, 4, 7, 11, \ldots$).

All right. A logical next question to ask is, how many pieces of pizza do you get when $C = 100$? My heavens! There must be a better way to spend your time than to run through the entire 100-step sequence. We need to be very clever.

Let us try the following. You will observe in the table that P increases faster than C but not as fast as C^2. Well, let's assume that a simple polynomial equation in C will help us determine the pattern of the P sequence. So we write

$$P = a_1 C^2 + a_2 C + a_3, \tag{3.1}$$

with the idea that we can determine the values of a_1, a_2, and a_3 with what is already known. So we "point match" at ($C = 0$, $P = 1$), ($C = 1$, $P = 2$), and ($C = 2$, $P = 4$). Substitution of these pairs of (C, P) values in equation (3.1) provides three equations containing the three unknowns a_1, a_2, and a_3. Some easy algebra gives $a_1 = 1/2$, $a_2 = 1/2$, and $a_3 = 1$. Accordingly, equation (3.1) becomes

$$P_2 = \frac{1}{2}(C^2 + C + 2). \tag{3.2}$$

The subscript 2 connotes the $n = 2$ case. If $C = 10$ then $P = 56$ and, in response to the question, if $C = 100$ then $P = 5,051$.

This problem is a portion of one of the *100 Great Problems of Elementary Mathematics* presented in the delightful book by Dörrie (1965). It and its somewhat more complicated extension—which we shall examine in a moment—were posed and solved by Jakob Steiner (1796–1863), a German mathematician who is considered to be one of the greatest of the modern geometers.

The problem, as stated by Steiner, is *What is the maximum number of parts into which a plane can be divided by C straight lines?* The answer given by Steiner is equation (3.2). For reasons that become clear shortly, it is written in the form

$$P_2 = 1 + C + \frac{1}{2}C(C - 1). \tag{3.3}$$

The subscript 2 is attached to emphasize that our pizza is a plane and so we are dealing with $n = 2$ dimensions.

An additional point needs to be mentioned. Frequently, we can describe a particular sequence in terms of a *recurrence relationship*. For example, in a later chapter we shall see that the sequence of the famous Fibonacci numbers can be specified by such a recurrence equation.

It turns out that we can also express our slicing number sequence by a recurrence relationship. A simple geometrical model yields the expression

$$P_{C+1} = P_C + 1 + C, \, n = 2, \tag{3.4}$$

where P_C and P_{C+1} are, respectively, the number of pieces after C and $C + 1$ slices. This result is easy to verify. Clearly, equation

(3.3) gives the number of pieces after C slices; call this P_C. Next, we replace C in equation (3.3) with the quantity $C + 1$. This gives the number after $C + 1$ slices; that is, P_{C+1}. Subtracting P_C from P_{C+1} gives the result of equation (3.4).

Backing Up to a Dot and a Licorice String

Two not very fierce problems are now quickly settled. The first is the $n = 0$ dimension case: a point or dot on a piece of paper. There is not much to do here since we do not slice dots. Accordingly, $P_0 = 1$.

The second is the $n = 1$ dimension case: a licorice string. If $C = 0$ then $P = 1$; if $C = 1$ then $P = 2$; if $C = 2$ then $P = 3$, and so on. So now we have the two equations

$$P_0 = 1; \; P_1 = 1 + C. \tag{3.5}$$

These expressions handle the $n = 0$ and $n = 1$ cases.

Now We Slice a Watermelon

With a close look at equations (3.3) and (3.5), perhaps we can see how a general solution is developing. But we need not be hasty. For the three-dimensional case ($n = 3$), we slice our watermelon once and so now we have two pieces, that is, $C = 1$, $P = 2$. Brilliant. Slice it again and we have $C = 2$, $P = 4$, and if $C = 3$ then $P = 8$. From here on, only those with unusually good spatial perception can deduce that if $C = 4$ then $P = 15$ and if $C = 5$ then $P = 26$.

Who needs spatial perception? Let mathematics do the work! We proceed with our analysis along two routes to obtain the solution we are after. We shall examine both of them.

Route 1

We construct a table similar to table 3.1 but extended to include the third differences, $\Delta^3 P$. We start the construction of table 3.2 with the $\Delta^3 P$ column and, working from right to left,

TABLE 3.2

The three-dimensional case ($n = 3$)

C	P	ΔP	$\Delta^2 P$	$\Delta^3 P$
0	1			
		1		
1	2		1	
		2		1
2	4		2	
		4		1
3	8		3	
		7		1
4	15		4	
		11		1
5	26		5	
		16		1
6	42			

obtain the indicated P column sequence. This procedure is based simply on the conjecture that mathematics is logical and orderly. If that is the case, we now know that the $n = 3$ sequence follows the pattern $1, 2, 4, 8, 15, 26, 42, 64, \ldots$

Route 2

As we did in the $n = 2$ case with equation (3.1), we write

$$P_3 = a_1 C^3 + a_2 C^2 + a_3 C + a_4, \tag{3.6}$$

in which our polynomial in C commences with C^3 instead of C^2. This time we point match with the first four (C, P) values shown in table 3.2. Substituting these into equation (3.6), we determine that $a_1 = 1/6$, $a_2 = 0$, $a_3 = 5/6$, and $a_4 = 1$. So the solution is

$$P_3 = \frac{1}{6}(C^3 + 5C + 6). \tag{3.7}$$

This is the answer provided by Dörrie (1965) to the problem presented by Steiner: What is the maximum number of parts into which a space can be divided by C planes?

Sure enough, the P values listed in table 3.2 agree with those computed from equation (3.7). Furthermore, for this $n = 3$ case, if $C = 10$ then $P = 176$ and if $C = 100$ then $P = 166,751$.

As before, we rewrite equation (3.7) in the form

$$P_3 = 1 + C + \frac{1}{2}C(C - 1) + \frac{1}{6}C(C - 1)(C - 2). \quad (3.8)$$

Again, we can construct the recurrence relationship for this three-dimensional case $(n = 3)$ in the same way we did for the two-dimensional case $(n = 2)$ that led to equation (3.4). The answer is

$$P_{C+1} = P_C + 1 + C + \frac{1}{2}C(C - 1), \; n = 3. \quad (3.9)$$

Slicing Stuff in the Fourth and Higher Dimensions

Good gracious! It looks as if the general case solution is staring right at us. In equation (3.8), P_0 is the first term, P_1 is the first two terms, P_2 is the first three terms, and P_3 is all four terms. Logically, we next seek the answer for P_4, that is, $n = 4$. To exactly paraphrase and extend Steiner's question: *What is the maximum number of parts into which a gizmo can be divided by C spaces?*

A gizmo is a piece of something in four-dimensional space; better terminology is suggested, perhaps the word *hypercube*. However, this does not help much when we get to $n = 5$ and beyond. Some thoughts about "four-dimensional intuition" are presented by Davis and Hersh (1981). In any event, we should not let our inability to visualize further than $n = 3$ prevent us from proceeding with analysis. Going on to higher dimensions is certainly no problem as far as mathematics is concerned.

As before, there are the two routes to the solution. In the first route we construct a table similar to tables 3.1 and 3.2 but, this time, we include a $\Delta^4 P$ column; once again, this consists of the sequence of ones. Working from right to left we obtain the $n = 4$ sequence: $1, 2, 4, 8, 16, 31, 57, 99, \ldots$

In the second route to the solution, we write down an equation, similar to equations (3.1) and (3.6). For this case, however, we surmise that we need to start with a C^4 term. That is,

$$P_4 = a_1 C^4 + a_2 C^3 + a_3 C^2 + a_4 C + a_5. \tag{3.10}$$

So now it is necessary to point match at five (C, P) values— (0, 1), (1, 2), (2, 4), (3, 8), (4, 16)—to determine the five coefficients; clearly, this will involve five simultaneous equations. The interesting book *The Lore of Large Numbers*, by Davis (1961), describes a procedure called the *method of successive elimination* to handle this kind of problem. An alternative approach is the *method of determinants*. Either methodology, with some algebra and arithmetic, provides the answer:

$$P_4 = \frac{1}{24}(C^4 - 2C^3 + 11C^2 + 14C + 24), \tag{3.11}$$

or, in the alternative form,

$$P_4 = 1 + C + \frac{1}{2}C(C - 1) + \frac{1}{6}C(C - 1)(C - 2)$$

$$+ \frac{1}{24}C(C - 1)(C - 2)(C - 3). \tag{3.12}$$

Similar methodology for the $n = 5$ case provides

$$P_5 = \frac{1}{120}(C^5 - 5C^4 + 25C^3 + 5C^2 + 94C + 120), \tag{3.13}$$

and for the $n = 6$ case,

$$P_6 = \frac{1}{720}(C^6 - 9C^5 + 55C^4 - 75C^3 + 304C^2$$

$$+ 444C + 720). \tag{3.14}$$

From here on, it is not worthwhile to obtain solutions in this "power series" form, since now we believe we know how to express the answer for any value of n. That is, looking at equation (3.8) (for $n = 3$) and equation (3.12) (for $n = 4$), we write, for $n = 5$,

$$P_5 = 1 + C + \frac{1}{2}C(C - 1) + \frac{1}{6}C(C - 1)(C - 2)$$

$$+ \frac{1}{24}C(C - 1)(C - 2)(C - 3)$$

$$+ \frac{1}{120}C(C - 1)(C - 2)(C - 3)(C - 4). \tag{3.15}$$

This seems to be the logical pattern for the P sequence. We surmise that the numbers appearing in the denominators of equation (3.15), that is, $1, 2, 6, 24, 120$, are in fact the *factorial* numbers. That is, for example, $4! = 4 \times 3 \times 2 \times 1 = 24$ and $5! = 5 \times 4 \times 3 \times 2 \times 1 = 120$.

Now on to another matter. If you would like some good exercise in algebra, you can show that equation (3.15) reduces to equation (3.13) for the $n = 5$ case. Further, you can easily write down the equation for the $n = 6$ case in a form similar to equation (3.15). Then show that this expression reduces to equation (3.14).

Finally, it is reasonable to infer that the general solution to our problem is

$$P_n = \sum_{j=0}^{n} \binom{C}{j}, \quad C \geq n, \tag{3.16}$$

where Σ is a symbol which indicates the summation of a series and

$$\binom{C}{j} = \frac{1}{j!}C(C - 1) \cdots (C - j + 1). \tag{3.17}$$

In addition, we have the special cases

$$\binom{C}{0} = 1; \quad \binom{C}{C} = 1. \tag{3.18}$$

For the limiting case in which $C = n$, Jolley (1961) gives

$$P_n = \sum_{j=0}^{C} \binom{C}{j} = 2^C, \; C = n. \qquad (3.19)$$

Expressions similar in form to equation (3.19) arise frequently in fields of mathematics called *probability theory* and *combinatorics*. In the vernacular of these fields, we would say that equation (3.17) gives the number of combinations of C things taken j at a time.

Some Examples of Slicing

So much for our analysis of the problem. Now for a few examples of applications.

EXAMPLE 1. A greedy land developer wants to know the maximum number of house lots he can obtain if he puts in 25 straight sidewalks across his large property. In this two-dimensional case ($n = 2$) we have from equation (3.16)

$$P_2 = \binom{C}{0} + \binom{C}{1} + \binom{C}{2},$$

which becomes, using equation (3.17),

$$P_2 = 1 + C + \frac{1}{2}C(C - 1).$$

So, with $C = 25$, we determine that $P = 326$ lots. The same answer is provided, of course, by equation (3.2).

EXAMPLE 2. A weird architect wants to acquire the maximum number of rooms in a very large tall building he is designing. How many rooms can he get if 10 planes — defining rather strange walls, ceilings, and floors — are passed through the building? In this case, $C = 10$ and $n = 3$. Using equations (3.16) and (3.17), we obtain

$$P_3 = \binom{10}{0} + \binom{10}{1} + \binom{10}{2} + \binom{10}{3}$$
$$= 1 + 10 + 45 + 120 = 176.$$

Consequently, the architect can have 176 rooms of various shapes and sizes.

EXAMPLE 3. How many super-gizmos can we obtain from ten-dimensional space with 10 slices? Here, $C = 10$ and $n = 10$. Since $C = n$, we can use equation (3.19) to determine that $P = 2^{10} = 1,024$.

Linear Pattern Analysis in Geography

An important problem in geography and regional science has to do with the quantitative description of linear patterns in geographical settings. These patterns may represent roads and railroads, boundaries of cities and towns, river and stream networks, political districts, school zones, and other features of physical geography. Various techniques have been developed by researchers in these disciplines to analyze and classify these geographical patterns.

As described by Coffey (1981), one of the techniques to examine linear patterns was developed by Dacey, who utilized two-dimensional *spacing* as the basis of discrimination among patterns. He studied the three linear spacing patterns shown in figure 3.1.

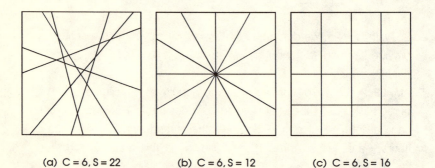

    (a) $C = 6, S = 22$ (b) $C = 6, S = 12$ (c) $C = 6, S = 16$

FIG. 3.1

Linear spacing patterns used in geographical analysis. (a) random, (b) radial, and (c) rectangular spacing.

Pattern 3.1(a) is essentially the $n = 2$ case we have been studying. This configuration is a random pattern of straight lines generated by connecting random points in a horizontal (x, y) plane. The equations of the array of lines are

$$y = a_i x + b_i, \tag{3.20}$$

where a_i and b_i are random values of slope and intercept.

If these quantities, a_i and b_i, are assigned designated values, then the random array of figure 3.1(a) reduces to the radial pattern of figure 3.1(b), at one extreme, and to the rectangular grid of figure 3.1(c), at the other. Detailed examinations of these geometries were carried out by Dacey; this enabled him to conduct so-called *nearest-neighbor* analyses. Such information is useful to geographers in studying things like marketing patterns, product distribution, traffic flow, settlement configuration, land utilization, and innovation diffusion.

We conclude with the observation that if figure 3.1 represents birthday cakes and if—for certain unspecified reasons—we are allowed exactly six slices, then obviously we get the most pieces of cake from figure 3.1(a). A bit messy and troublesome to serve perhaps, but still the most pieces.

4

*Raindrops Keep Falling
on My Head
and Other Goodies*

Which weighs more, all of the air in the world or all the water, and how much more? What would happen if all the ice in the world were to melt? What is the average rainfall in the world? What is the velocity and kinetic energy of a raindrop? How much power is in rainfalls? How many times has the world's air been breathed by humans and the world's water drunk by humans?

We now obtain answers to these awesome questions. Nothing difficult in the way of theoretical analysis is introduced. However, there will be quite a few numbers and calculations to consider. We shall be using information given in the text on oceanography by Sverdrup, Johnson, and Fleming (1942) and the intriguing book on environmental science by Harte (1985).

Amount of Air in the World

It is fairly easy to determine how much air there is in the world. We know that at sea level the atmospheric pressure is $p_0 = 1.01325 \times 10^5$ newtons/m^2. By definition, this amount is simply the weight of the column of air acting on each square meter of the earth's surface. If we divide this value by the gravitational acceleration, $g = 9.82$ m/s^2, we obtain $p_0 =$

10,318 kg/m^2. This is the mass of air resting on every square meter of the earth.

Well, how many square meters are there? The average radius of the earth is $R = 6.37 \times 10^6$ m and the area of a sphere is $A = 4\pi R^2$. So we determine that $A = 5.10 \times 10^{14}$ m^2 is the area of the earth.

With the above information, the total weight of the air—or, more precisely, the total mass of the atmosphere—is

$$M_{air} = p_0 A = (10,318)(5.10 \times 10^{14}) = 5.262 \times 10^{18} \text{ kg.}$$

This answer can be improved slightly. Recall that p_0 is the pressure at sea level. However (and thank goodness), not all of the earth's surface is at sea level. According to Harte (1985), the average elevation of the continents is $z = 840$ m above sea level. In addition, from Harte, we know that the area of the oceans is $A_0 = 3.61 \times 10^{14}$ m^2 (70.8% of the total) and the area of the continents is $A_c = 1.49 \times 10^{14}$ m^2 (29.2%).

Since the average elevation of the land surface is 840 meters above sea level, it is clear that the atmospheric pressure is less than at sea level and, in turn, the mass of the air column is less.

It is possible to calculate the atmospheric pressure at a specified elevation by substituting the so-called general gas law into the equation of fluid statics. Doing so, we obtain the expression

$$p = p_0 e^{-(mg/R_* T)z}, \tag{4.1}$$

in which p_0 is the atmospheric pressure at sea level and p is the atmospheric pressure at any elevation, z, above sea level. The quantity e is the base of natural logarithms; it has the numerical value $e = 2.71828$. Later on we shall examine this famous number, e, to quite an extent. The other quantities in equation (4.1) are

R_*: gas constant; $R_* = 8.314$ joule/°K mol

m: molecular weight of air; $m = 28.96 \times 10^{-3}$ kg/mol

g: gravitational acceleration; $g = 9.82$ m/s^2

T: absolute temperature; T (°K) = °C + 273°; $T = 15° + 273° = 288$ °K

Substitution of these numerical values into equation (4.1), along with $z = 840$ m, gives $p/p_0 = 0.905$.

This result indicates that at a temperature of 15°C and an elevation of 840 meters, the atmospheric pressure is 90.5% of the pressure at sea level. Consequently, the mass of air acting on each square meter of land surface is 9,338 kg. By the way, to simplify the analysis leading to the result of equation (4.1), we had to assume an isothermal (constant temperature) atmosphere. Although this is not a very realistic assumption, equation (4.1) is surprisingly accurate for elevations up to 3,000 meters.

Now we are able to compute the total mass of air somewhat more precisely:

$$M_{\text{air:ocean}} = (10,318)(3.61 \times 10^{14}) = 3.725 \times 10^{18} \text{ kg,}$$

$$M_{\text{air:land}} = (9,338)(1.49 \times 10^{14}) = 1.391 \times 10^{18} \text{ kg,}$$

So

$$M_{\text{air}} = 5.116 \times 10^{18} \text{ kg.}$$

Amount of Water in the World

There is no way we can calculate how much water there is in the world; we simply have to look it up. Table 4.1 lists this information; it gives the volumes of water according to type of reservoir.

We note that the oceans hold over 97% of the earth's water. The Pacific contains 52.8% of the ocean total, followed by the Atlantic (including the Arctic) with 25.9% and the Indian with 21.3%.

The total volume of the five Great Lakes of North America is 0.0227×10^{15} m^3. Together they contain about 18% of the world's surface fresh water.

From table 4.1 and the formula $M = \rho V$, where ρ is the density, we easily compute the total mass of water in the world:

$$M_{\text{water:salt}} = (1,035)(1,350 \times 10^{15}) = 1.397 \times 10^{21} \text{ kg,}$$

$$M_{\text{water:fresh}} = (1,000)(37.5 \times 10^{15}) = 0.038 \times 10^{21} \text{ kg,}$$

TABLE 4.1

Volume of water in the world

Type of reservoir	Volume, V $10^{15}\ m^3$	Percentage of total
Oceans	1,350	97.30
Ice caps and glaciers	29	2.09
Ground water and soil	8.37	0.60
Lakes, rivers, and streams	0.125	0.009
Atmosphere	0.013	0.001
Total	1,387.5	100.00

Source: Harte (1985).

So

$$M_{water} = 1.435 \times 10^{21} \text{ kg}.$$

We now summarize the results of this brief analysis. The ratio of the mass of water to the mass of air is 1.435×10^{21} kg/ $(5.116 \times 10^{18}$ kg$) = 280$. So the answer to the question at the beginning of this chapter is that all of the water in the world weighs 280 times more than all the air in the world. This is an interesting result. However, an even more interesting aspect of this short problem is that we were able to determine the two total weights, water and air, and their ratio, without much uncertainty and with relatively little difficulty.

Suppose that All the Ice Were to Melt

Here is an interesting question to examine. What would happen if all the ice presently locked up mostly in the polar ice caps and glaciers were to melt? How much would the surface of the ocean rise? How much of the present land areas of the world would be flooded?

From table 4.1 it is seen that the volume of water contained in the ice caps and glaciers is $V_{ice} = 29 \times 10^{15}$ m^3. The surface area of the oceans is $A_0 = 3.61 \times 10^{14}$ m^2. Dividing the first of these quantities by the second gives the increase in depth, $z = 80.3$ m.

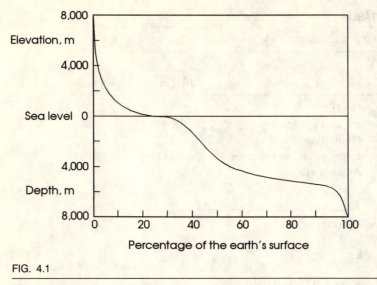

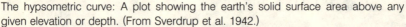

FIG. 4.1

The hypsometric curve: A plot showing the earth's solid surface area above any given elevation or depth. (From Sverdrup et al. 1942.)

However, as the ocean depth increases, due to melting ice, the ocean area will enlarge as water begins to flood the low-elevation regions of the world. Consequently, the increase in ocean depth will be something less than 80 meters.

To answer the questions we need the so-called hypsometric (or hypsographic) curve shown in figure 4.1. This curve shows the percentage of the earth's solid surface above any given depth or elevation. For example, 29.2% of the earth's land surface is above $z = 0$ (sea level), 7.9% is above 1,000 meters, and so on.

We now construct a simple mathematical model to guide our analysis. A moment ago we established that the increase in ocean depth due to the melted ice will be something less than 80 meters; how much less is the problem. To solve the problem, it is clear that we need an equation of the hypsometric curve shown in figure 4.1 for values of the elevation z less than, say, 100 meters or so.

As precisely as we can, we match or fit the hypsometric curve of figure 4.1 with a simple straight line relationship applicable for small values of z. Doing so yields the following result:

$$z = 17.1(29.2 - P_{land}), \tag{4.2}$$

in which land P_{land} is the area of the earth's land surface as a percentage of the earth's total area. Thus, if $z = 0$ (i.e., present sea level) in equation (4.2), we obtain $P_{land} = 29.2\%$, a result we already have. If $z = 100$ m, for example, then $P_{land} = 23.4\%$.

Next, using the relationship $P_{land} + P_{ocean} = 100.0\%$, we alter equation (4.2) to the form,

$$z = 17.1(P_{ocean} - 70.8), \tag{4.3}$$

where P_{ocean} is the area of the earth's ocean surface as a percentage of the earth's total area.

Finally, knowing that the earth's total area is $A_T = 5.10 \times 10^{14}$ m^2, we obtain the expression,

$$z = 335 \times 10^{-14}(A - A_0), \tag{4.4}$$

in which A is the area of the ocean surface when the surface is at elevation z and $A_0 = 3.61 \times 10^{14}$ m^2 is the area when $z = 0$. Solving this equation for A, we obtain

$$A = A_0 + (0.0030 \times 10^{-14})z. \tag{4.5}$$

Now we are ready for the final step. From table 4.1, the volume of the ice caps and glaciers is $V = 29 \times 10^{15}$ m^3. Since all this ice melts into the ocean, we have

$$29 \times 10^{15} = z[(3.61 \times 10^{14}) + (0.0030 \times 10^{14})z]$$

or

$$290 = 3.61z + 0.0030z^2. \tag{4.6}$$

This is a quadratic equation in z. Recall that if the quadratic equation is written in the general form,

$$ax^2 + bx + c = 0, \tag{4.7}$$

then its solution is given by

$$x = \frac{1}{2a}\left[-b \pm \sqrt{b^2 - 4ac}\right].$$ (4.8)

Using this expression, the solution to equation (4.6) is $z = 75$ m. So we have our answer. If all the world's ice were to melt, the oceans would rise by about 75 meters (246 feet).

Finally, we note that the quantity $A - A_0$ represents the area of land surface flooded by the rise of the ocean. We easily compute from equation (4.5) that the flooded area is 0.225×10^{14} m^2 or approximately 22.5 million km^2. The area of the contiguous 48 states of the United States is 7.62 million km^2. So we conclude that the total area flooded by the melted ice, worldwide, would be about three times larger than the 48 states of the United States.

A SUGGESTION FOR A TERM PAPER. Are you looking for an interesting subject for a term paper in your course in geography, science, or mathematics? Well, you might want to use this topic of ice melting in the oceans, but explore the matter to a much greater extent. A good place to begin your analysis is the interesting book by de Blij and Muller (1996).

For example, let's assume that the ocean level rises by 75 meters due to the melted ice. In your term paper, you could examine and respond to the following questions:

What would the substantially altered map of North America look like? What would be the new area of the forty-eight states of the United States? What major cities would be submerged?

Most of Florida vanishes. What about the other low-elevation states, such as Delaware, Louisiana, Mississippi, and Rhode Island? How much of Hawaii disappears?

How are the Saint Lawrence River Valley, Niagara Falls, and the Great Lakes affected? What happens to the Panama Canal?

How are the Central Valley, Death Valley, and Imperial Valley in California changed? Where are the mouths of the Colorado, Mississippi, Columbia, Rio Grande, and Hudson Rivers?

In the United States, approximately how many people have to move
to higher ground?

Some Items Involving the Hydrologic Cycle

Most of us have heard about the so-called water cycle or
hydrologic cycle. This is the process featuring evaporation of
water from ocean and land surfaces, retention of this water vapor
in the atmosphere for a relatively short period of time, precipitation of this water, usually in the form of rain, onto oceans and
lands, and retention of the water in the ocean, ground, lakes, and
rivers for a relatively long period of time. An important compo-
nent of the cycle is the drainage or runoff of water from land
surfaces to the oceans. A diagram illustrating the hydrologic cycle
is shown in figure 4.2.

First, we consider evaporation. As indicated in the figure, the
total rate of evaporation is $Q = 456 + 62 = 518 \times 10^{12}$ m^3/yr.
The rate from the surface of the oceans is given by the equation
$q = Q/A_0 = (456 \times 10^{12})/(3.61 \times 10^{14}) = 1.263$ m/yr, or about
49.7 inches per year. From land surfaces the rate is 0.416 m/yr or

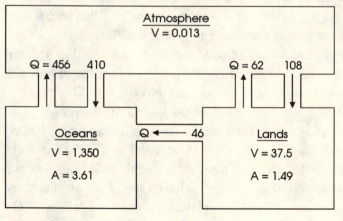

FIG. 4.2

The hydrologic cycle. Units of the quantities are (a) V: 10^{15} m^3, (b) A: 10^{14} m^2, and
(c) Q: 10^{12} m^3/yr.

16.4 inches per year. Worldwide, the average rate of evaporation is $q = 1.016$ m/yr = 40.0 in/yr.

Next, the matter of average retention time or residence time in the atmosphere is examined. Suppose we have a tank containing $V = 30$ gallons and that water flows into and out of the tank at the rate $Q = 5$ gallons per minute. Then the average residence time of a water particle in the tank is $T = V/Q = 6$ minutes.

By the same token, from figure 4.2 it is noted that the amount of water stored in the atmosphere is $V = 0.013 \times 10^{15}$ m^3, and the rates of inflow (evaporation) and outflow (precipitation) are $Q = 518 \times 10^{12}$ m^3/yr. Accordingly, the average residence time of a water molecule in the atmosphere is $T = 0.025$ yr = 9.2 days.

Now we go to the topic of precipitation or rainfall. From the figure we see that the total rate of precipitation is $Q = 410 + 108 = 518 \times 10^{12}$ m^3/yr. The average rainfall rate over the oceans is $q = 1.136$ m/yr = 44.7 inches per year and over land surfaces $q = 0.725$ m/yr = 28.5 inches per year.

On a worldwide basis, the average rainfall is 1.016 m/yr or 40.0 inches per year. Needless to say, there is a wide range of rainfall rates. Rainfalls of nearly 500 inches per year have been recorded in Hawaii and India. At the other extreme, in desert regions in various parts of the world there is virtually no rainfall whatsoever. The Atacama desert in northern Chile holds the record, with an average annual rainfall of 0.02 inches.

Returning to the matter of residence time, we saw that a water molecule is retained in the atmosphere for an average of only 9.2 days between the time it is evaporated and the time it is precipitated. We make the same kind of computation of average residence time for a water molecule on the earth's surface. That is, $T = V/Q = (1{,}387.5 \times 10^{15})/(518 \times 10^{12}) = 2{,}679$ yr. This means that, on the average, a water molecule falling in 680 BC. was finally evaporated in 1999 in order to spend about nine days in the atmosphere before falling back to earth once again. Of course, the residence time of 2,679 years is the average. The residence time of a particular molecule could be a few minutes or it could be many thousands of years.

Finally, a word about runoff. The water balance diagram of figure 2.4 shows a drainage of $Q = 46 \times 10^{12}$ m^3/yr from the lands to the oceans. This represents the combined discharges of all the world's rivers, streams, and underground aquifers into the Pacific, Atlantic, Indian, and Arctic oceans. The annual flow indicated above is equivalent to $Q = 1,460,000$ m^3/s. To give a scale to this quantity, we note that the world's mightest river, the Amazon, has an average discharge at its mouth of about 180,000 m^3/s. So the Amazon alone handles around 12% of all the world's runoff to the seas.

Our thrilling saga will continue after a brief pause.

5

Raindrops and Other Goodies Revisited

Runoff of Rivers

Our saga now continues. Readers will recall that the incredible Amazon river in South America delivers 180,000 cubic meters of water per second to the Atlantic ocean. This is about thirty times more than the flow over Niagara Falls. The drainage area of the Amazon is about 7 million square kilometers—about ten times the size of Texas. Much of the Amazon basin receives heavy rainfall to sustain the well-known rain forests.

On a smaller scale, the same things can be said about the world's second river: the Congo (or Zaire) in Africa. It collects water from a drainage area of 3.5 million square kilometers and empties over 41,000 cubic meters per second into the Atlantic. The Congo basin is also a region of tropical rain forests and extensive rainfall.

And so it goes. The fifteen major rivers of the world (scaled on the basis of discharge, not length), collectively drain about 37 million square kilometers (24.7% of the world's land area). Together, they deliver 460,000 cubic meters of water per second to the oceans (31.5% of total runoff). Eight of these rivers are in Asia, three each in North America and South America, and one in Africa. They are listed in table 5.1.

TABLE 5.1

Average discharges of rivers (10^3 m^3/s)

Amazon	180	Orinoco	20	Irrawaddy	13
Congo	41	Yenisei	19	Amur	12
Brahmaputra	38	Mississippi	18	Mekong	11
Yangtze	34	Lena	16	Mackenzie	11
La Plata	22	Ob	15	St. Lawrence	10

Kinetic Energy and Power of Rainfalls

Enough of rivers; back to rainfalls. Not surprisingly, in rainfalls there are various sizes of raindrops. For example, in a light shower ($r = 5$ mm/hr), raindrop diameters range from about $D = 0.1$ mm to $D = 1.5$ mm. In a moderate rainfall ($r = 25$ mm/hr), the maximum diameter is around $D = 3.5$ mm. And in a heavy deluge ($r = 100$ mm/hr), drop diameters can be as much as $D = 6.0$ mm.

Maximum raindrop diameter is about 6 millimeters. Larger sizes tend to be unstable due to shearing forces caused by their motion through the air. Consequently, they rupture and form smaller drops.

What is the velocity of a raindrop? As we shall see, it depends on the diameter of the drop. The raindrop velocity also depends on the elevation of the rainfall above sea level. It is logical that raindrops fall faster at higher elevations because the air is less dense. If you are interested in the complexities of raindrops, and indeed there are a great many, texts dealing with atmospheric science and cloud physics are good places to start. Suggested references are the books on fluid mechanics by Richardson (1952) and on environmental aerodynamics by Scorer (1978).

For our present purpose, a simple equation to compute the terminal velocity of a raindrop is the expression provided by Dingle and Lee (1972):

$$U = 0.0549D^3 - 0.888D^2 + 4.9184D - 0.166, \qquad (5.1)$$

where D is the drop diameter in millimeters and U is the velocity in meters per second. Thus, if $D = 2.5$ mm, we obtain $U = 7.44$ m/s.

We have now set the stage for some very exciting calculations. As we saw before, the earth receives a total of 518×10^{12} cubic meters of precipitation each year, mostly as rain but some as snow and hail. Here's one for people who like big numbers: how many raindrops are contained in this 518 million million cubic meters of annual rainfall?

To make things easy, suppose all the annual rainfall were composed entirely of raindrops with diameter $D = 3$ mm. The volume of a sphere is $v = \pi D^3/6 = \pi(3)^3/6 = 14.137$ mm^3 or $v = 14.137 \times 10^{-9}$ m^3. So the number of drops falling each year is

$$n = \frac{Q}{v} = \frac{518 \times 10^{12}}{14.137 \times 10^{-9}} = 3.664 \times 10^{22}.$$

This is quite a few raindrops. To carry this kind of numerical computation a step further, please verify that with this many drops, every square millimeter of the earth is hit, on the average, by 72 raindrops every year. A good thing to know.

The kinetic energy of a single raindrop is

$$e = \frac{1}{2}mU^2 = \frac{1}{2}\rho v U^2. \tag{5.2}$$

Since the diameter of our drop is $D = 3$ mm, its velocity, from equation (5.1), is $U = 8.08$ m/s. With density $\rho = 1,000$ kg/m^3 and volume $v = 14.137 \times 10^{-9}$ m^3, we compute from equation (5.2) that $e = 4.615 \times 10^{-4}$ joule.

For a single raindrop this is certainly not much energy. But we have a lot of drops. The total power produced by all of the raindrops is

$$P = en = 16.91 \times 10^{18} \frac{\text{joule}}{\text{yr}} = 536 \times 10^6 \text{ kW}.$$

This is the result we are after. Now if somehow the kinetic energy of the world's rainfalls could be captured, we would be able to generate 536 million kilowatts of power. To provide a scale to this number: in 1990 the total power-generating capacity of the United States (hydroelectric and steam) was 698 million kilowatts.

Of course, it is not feasible to harness the kinetic energy of raindrops. However, we can appreciate how much power there is in rainfalls. Admittedly, rainfalls do help to alleviate pollution in rivers, lakes, and estuaries by (a) direct addition of oxygen from the oxygen-saturated drops, (b) increasing the rate of oxygen transfer from the atmosphere to the body of water, and (c) outright dilution ("the solution to pollution is dilution"). Otherwise, the enormous kinetic energies of rainfalls do little beyond causing extremely serious problems of soil erosion.

At this point, you might be interested in confirming that rainfalls comprised only of $D = 2$ mm drops will generate 354 million kW and rainfalls with $D = 4$ mm drops will produce 637 million kW.

How Many Times Has the World's Water Been Consumed by Humans?

As we have seen, there is an enormous amount of water in the world. Furthermore, we know that a great many people live and have lived on our planet in the past. So a reasonable and interesting question is the following: how many times have all these humans consumed all this water during the past million years or so?

First we turn to the extensive *Scientific Tables* prepared by Ciba-Geigy and edited by Diem and Lentner (1970). They indicate that per capita daily water intake in all forms (liquids and food) is approximately 2.5 liters. Accordingly, water consumption is $M = 2.5$ kg/day = 912.5 kg/yr per person.

In chapter 15, "How Many People Have Ever Lived?," we establish that about 80 billion people lived on earth during the period from 1,000,000 B.C. to 1990. With an assumed average life span of 25 years, we are dealing with a total of 2,000 billion person-years. Consequently, the total amount of water consumed by humans during the past million years or so is M(water consumed) $= (912.5)(2{,}000 \times 10^9) = 1.825 \times 10^{15}$ kg. The amount of water in the world is M(water available) $= 1.435 \times 10^{21}$ kg. Dividing the first quantity by the second gives 1.27×10^{-6}. This means that humans have consumed only about one cup of water per million cups available. Or, stated in a more direct fashion, over the past million years humans have consumed only about one-twelfth of the volume of the Great Lakes; in other words, about the same amount as Lake Ontario.

How Many Times Has the World's Air Been Breathed by Humans?

Our final question has a rather startling answer. The *Scientific Tables* of Diem and Lentner report that human breathing rates range from about 1.0 liters per minute for infants to around 40 liters per minute for men engaged in heavy work. A reasonable average value is $Q = 12 \, \text{l/min} = 0.2 \times 10^{-3} \, \text{m}^3/\text{s}$. Next, we want to calculate the mass of air we breathe per second. Now the density of air (sea-level pressure; 15°C) is $\rho = 1.225 \, \text{kg/m}^3$. Using the formula $M = \rho Q$, we determine that $M = 0.245 \times 10^{-3}$ kg/s $= 7{,}725$ kg/yr per person.

As before, from the year 1,000,000 B.C. to 1990, there were 2,000 billion person-years of human life. So the computed amount of air breathed by humans during that very long period of time is M(air breathed) $= 1.56 \times 10^{16}$ kg.

The amount of air in the world is M(air available) $= 5.116 \times 10^{18}$ kg. The ratio of the amount of air breathed to the amount available is 0.00305 or 0.305%. This answer says that only about 0.3% of the air in the world has been breathed even once by humans. Could this be true? It seems to be a surprisingly small amount. Still, we have defined our computation and that is the outcome.

We conclude our chapter with the following interesting item: Annually, a person breathes about 7,700 kilograms of air and consumes around 900 kilograms of water, or about 8.5 times more air than water.

6

Which Major Rivers Flow Uphill?

As all of us know, jellied biscuits fall to the rug (frequently upside down), baseballs fall to the center fielder (frequently with two outs and bases loaded), and car keys fall to the pavement (frequently through sewer gratings). And rivers fall—or as we usually say—flow downhill. So how can a river flow uphill? Well, be patient. That matter will be explained in a page or two.

We start our consideration of this topic with a short trip through some trigonometry and analytic geometry. As illustrated in figure 6.1, suppose that you have a circular cone made of cheese on the table in front of you. With the blade of your knife or saw parallel to the table top, you slice through the cone of cheese. The cross-section of the cut cone is a *circle*. Your second

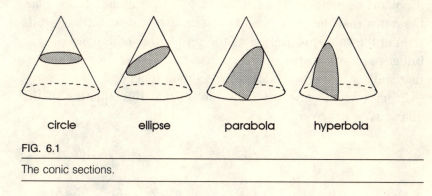

circle   ellipse   parabola   hyperbola

FIG. 6.1

The conic sections.

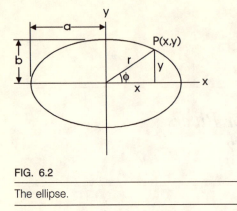

FIG. 6.2

The ellipse.

slice, with the knife blade held at a small angle to the plane of the table, yields an *ellipse*. The third slice, cut at an angle parallel to the side of the cone, gives a *parabola*. And your fourth slice, with the knife blade at a steep or even a vertical angle, produces a *hyperbola*.

These shapes—the circle, ellipse, parabola, and hyperbola—are the "conic sections." They have been studied and analyzed by mathematicians for more than two thousand years. Incidentally, if you do not happen to have a cone-shaped piece of cheese around, you can use a flashlight and a nearby wall to view the conic sections. This methodology is easier, less expensive, and involves fewer calories.

For our uphill-flowing river problem, we are interested in only the second of these conic sections: the ellipse shown in figure 6.2. This is an extremely important curve; it appears in a great many problems of mathematics and science. For example, it describes the paths that the planets and most comets follow in their orbits around the sun. It is utilized as the profile for some types of arch bridges and other structures. It provides the basis for an important subject in mathematics called elliptic functions.

From analytic geometry we establish that the equation of the ellipse is

$$\frac{x^2}{a^2} + \frac{y^2}{b^2} = 1, \qquad (6.1)$$

where x and y are the coordinates of a particular point on the curve, a is half the length of the major (longer) axis of the ellipse, and b is half the length of the minor (shorter) axis. In the special case $a = b = r_0$, equation (6.1) reduces to $x^2 + y^2 = r_0^2$; this is the equation of the circle. From equation (6.1) we obtain the expression

$$y = \pm \frac{b}{a}\sqrt{a^2 - x^2},\qquad (6.2)$$

which says that for every value of the quantity x, there are two corresponding values of the quantity y. Suppose that $a = 5$ and $b = 3$. For the value of $x = 4$, for example, we determine from equation (6.2) that $y = +1.8$ and -1.8.

Back to equation (6.1) and figure 6.2. If this ellipse is rotated about the *x-axis* we generate a solid of revolution that looks somewhat like a football or watermelon; it is called a *prolate ellipsoid*. Alternatively, if the ellipse is rotated about the *y-axis*, the solid of revolution is termed an *oblate ellipsoid*; it resembles a slightly flattened pumpkin or the planet earth.

That is correct: the planet earth. As we know, the earth makes a complete rotation about its axis every 24 hours. This rotation is sufficiently fast to produce a velocity of about 460 meters per second at the equator. Like everything else that rotates about an axis, the earth creates centrifugal forces. For the rotating earth, these forces are large enough to significantly distort its shape from a perfect sphere to a slightly flattened pumpkin, that is, an oblate ellipsoid.

According to Abramowitz and Stegun (1965), the *equatorial* radius of the earth is $a = 6,378,388$ m and the *polar* radius is $b = 6,356,912$ m. The difference between these two distances indicates that the earth's diameter measured between two points on the equator on opposite sides of the earth is about 43 kilometers greater than the diameter from the north pole to the south pole.

We are now at the threshold of our "river flowing uphill" problem. Indeed, here is the gimmick (a bit sneaky, but never mind): What major rivers of the world have their mouths further from the center of the earth than their sources? As you eagerly endeavor to find answers to the question, you will be forced to combine these topics of mathematics with some topics of geography.

First we cast our mathematics into a more useful form. Referring to figures 6.2 and 6.3 and the relationships

$$x = r \cos \phi, \ y = r \sin \phi, \tag{6.3}$$

where r is the distance to the center of the earth and ϕ is the latitude of a particular point on the surface, we convert equation (6.1) into the form

$$r = \frac{1}{\sqrt{\dfrac{\cos^2 \phi}{a^2} + \dfrac{\sin^2 \phi}{b^2}}}. \tag{6.4}$$

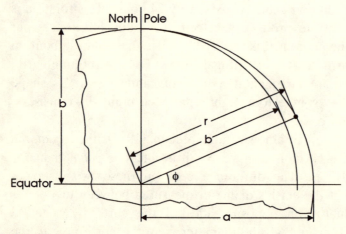

FIG. 6.3

Definition sketch for the coordinates of the earth considered as an oblate ellipsoid.

TABLE 6.1

Values of radial distance r and radial distortion $(r - b)$ for various values of latitude ϕ

ϕ degrees	r meters	$(r - b)$ meters
0	6,378,388	21,476
10	6,377,737	20,825
20	6,375,864	18,952
30	6,372,999	16,086
40	6,369,488	12,576
50	6,365,759	8,847
60	6,362,261	5,349
70	6,359,413	2,501
80	6,357,556	644
90	6,356,912	0

Since this is somewhat awkward to use for computation, we utilize the expression $\sin^2 \phi + \cos^2 \phi = 1$, to obtain

$$r = \frac{b}{\sqrt{1 - \left\{1 - \left(\frac{b}{a}\right)^2\right\}\cos^2 \phi}}. \tag{6.5}$$

Substituting $a = 6,378,388$ m and $b = 6,356,912$ m into this equation gives

$$r = \frac{6,356,912}{\sqrt{1 - 0.006722653\cos^2 \phi}}. \tag{6.6}$$

Table 6.1 lists values of r for various values of latitude, ϕ. It also indicates the amount of radial distortion $(r - b)$ due to centrifugal force.

Now what rivers are likely to be uphill flowing? First, we exclude those flowing mostly west to east or east to west, since, in these cases, there is little or no change in latitude. Evidently we seek north to south or south to north rivers. Obviously, you will need some maps and tables to proceed with your analysis. A good deal of information can be obtained from the excellent atlas of Rand McNally (1978). Contour maps are needed to determine elevations of sources.

A good first guess is: the Mississippi River. This river flows almost directly north to south. It has its source in Lake Itasca in northern Minnesota at a latitude $\phi_1 = 47.5°$, and an elevation $h_1 = 450$ m above mean sea level. From equation (6.6), the distance from the center of the earth to sea level at this latitude is $r_1 = 6,366,687$ m. The source is 450 meters higher. So $d_1 = r_1 + h_1 = 6,367,137$ m is the distance from the source of the Mississippi to the center of the earth.

The mouth of the Mississippi is about 120 kilometers southeast of New Orleans in the Gulf of Mexico at a latitude $\phi_2 = 29.0°$. At sea level we have $h_2 = 0$. From equation (6.6) we compute $r_2 = 6,373,321$ m, and since $h_2 = 0$, the distance from the mouth of the river to the center of the earth has the value $d_2 = r_2 + h_2 = 6,373,321$. The difference between these two distances is $\Delta d = d_2 - d_1 = +6,184$ m. This result indicates that the mouth of the Mississippi River is 6,184 meters further from the center of the earth than is its source. Accordingly, this major river "flows uphill."

The Missouri River, with its source in Madison County, Montana, and mouth in the Mississippi River near Saint Louis, also flows uphill. In this case, $\Delta d = 2,365$ m. The Rio Grande starts in San Juan County, Colorado, at an elevation $h_1 = 1,830$ m and latitude $\phi_1 = 37.7°$. Its mouth is in the Gulf of Mexico with $\phi_2 = 26.0°$ and $h_2 = 0$. So, with $\Delta d = 2,080$ m, the Rio Grande flows uphill. The Colorado-Green River also flows uphill with $\Delta d = 650$ m.

Are there other uphill-flowing rivers in North America? Undoubtedly. Try the Arkansas and the Ohio; possibly the Connecticut, Hudson, and Sacramento. However, the Columbia, Saint Lawrence, and Yukon Rivers, with small latitude changes, probably do not qualify. The Mackenzie River, though it flows mainly south to north, flows the wrong way to be a candidate for uphill flowing.

On this latter point, the Nile River flows almost due north from its source in Lake Victoria ($\phi_1 = 2.0°$, $h_1 = 1,135$ m) to its mouth in the Mediterranean ($\phi_2 = 31.5°$, $h_2 = 0$). Accordingly, $d_1 = 6,379,498$, $d_2 = 6,372,503$, and the difference $\Delta d = d_2 - d_1 = -6,995$ m. The minus sign means that the Nile is flowing downhill in two senses: (a) it goes from a high place to a low place on the earth's surface and (b) its mouth is *closer* to the earth's center than is its source. Too bad; the Nile does not even come close.

The important criterion to be "uphill flowing" should now be clear: the river must have its source at high latitude and its mouth at low latitude.

PROBLEMS As best you can, calculate Δd for some of these rivers: Amazon, Mekong, Volga, Yenisei, Danube, Rhine, Yangtse, and Zaire.

An Epilogue: The Shape of the Earth

Two comments: First, we have covered some topics involving *mathematics* and *geography* in our deliberations about rivers flowing uphill. Let's go a step further and include a bit of *history* in our considerations. Second, though we label this section an "epilogue" to our chapter, it is distinctly a "prologue" to the entire matter.

A word of explanation: As recently as the early eighteenth century, there was not even agreement among the world's scientists about the shape of the earth. The French believed it to be a prolate (watermelon-shaped) ellipsoid; the British believed it to be an oblate (pumpkin-shaped) ellipsoid.

In order to settle the matter, in the spring of 1735, the French Academy of Sciences organized two scientific expeditions to measure arcs of the earth's surface: one near the equator and the other as close as possible to the North Pole. Soon after, two teams were deployed, one to Ecuador in South America, the other to northern Sweden in Europe. Though the team to Sweden returned to France a year later, ten years passed before the Ecuadorian team finally got back to Paris.

An extremely interesting story of the experiences and main achievements of the two scientific expeditions is given by Fernie (1991, 1992) in a series of three fascinating articles entitled "The Shape of the Earth," which are highly recommended reading. By the way, as we all know, the earth is oblate.

A Brief Look at π, e, and
Some Other Famous Numbers

The professor went to the board one day
And asked his class, just for the fun
What is the only conceivable way
To combine e, i, π, zero and one?

A brilliant young scholar, he proved quite a hero
Who knew the professor just loved to tease
Replied: $e^{i\pi} + 1 = 0$
Then he requested: the next question, please.

Many years ago, the writer received a magnificent $15 for submitting this poem, which, not long afterward, appeared in the mathematical nursery rhyme corner of a trade journal. Though the poem is pathetic enough, the real misfortune is that space did not allow acknowledgment to the famous Swiss mathematician, Leonhard Euler. In any event, the episode serves as a prologue to our next endeavor: an examination of several of the important numerical constants of mathematics.

Without question, the most famous of these "numbers" is the one we call π; it has the approximate numerical value $\pi = 3.14159$. Although its basic definition relates the circumference of a circle to its diameter, π appears in a large number of mathematical problems that have nothing whatsoever to do with circles.

Another extremely important number is the one identified as *e*. Its approximate value is $e = 2.71828$. It serves as the basis for so-called natural logarithms and also for things like exponential growth in demography, radioactive decay in physics, and bell-shaped curves in probability theory. Another famous number is the one called the golden ratio, $\phi = 1.61803$. This number shows up in the strangest places, including the architecture of the Parthenon in Greece and the Great Pyramids of Egypt, the geometry of five-pointed stars and logarithmic spirals, and the shape of sunflower blossoms and Nautilus sea shells.

Yet another important number is Euler's constant. Its numerical value is $\gamma = 0.57721$. Again, this number appears in many problems of mathematics including the theory of heat conduction and the theory of extreme value distribution in statistics. Our fifth and final number is also the newest one: the Feigenbaum number, $\delta = 4.6692$. This numerical constant makes its appearance in the relatively new area of mathematics called chaos theory.

In addition to these *mathematical* constants, there are also a great many *physical* constants; more than thirty of them are listed by Abramowitz and Stegun (1965). The ones we use in our various analyses are the following:

Gravitational acceleration, $g = 9.82$ m/s^2 (in fact, g changes slightly with latitude, varying from 9.78 at the equator to 9.83 m/s^2 at the poles)

Gas constant, $R_* = 8.314$ joule/°K mol

Gravitational constant, $G = 6.673 \times 10^{-11}$ newton m^2/kg^2

Speed of light, $c = 2.998 \times 10^8$ m/s

The Most Famous Number of All: π

The most ancient, most familiar, and most important number in all of mathematics—indeed in all of human civilization—is the one we designate with the symbol π. It represents the ratio of the circumference of a circle and its diameter and has the approximate value $\pi = 3.14159$. Thus, if C and D are, respectively, the circumference and diameter of a circle, then $C = \pi D$. Also, if A is the area of a circle and $R = D/2$ is its radius, then $A = \pi R^2$.

More than 4,000 years ago, the Babylonians had established an approximate value for this important number: $\pi = 25/8$ (that is, 3.125). At about the same time, the Egyptians had determined that $\pi = 256/81$ (3.1605). The greatest mathematician of antiquity, Archimedes of Syracuse (287–212 B.C.), knew that π was more than 223/71 (3.14085) but less than 22/7 (3.14286).

Surprisingly accurate numerical values of π were also known in ancient Chinese, Hindu, and Mayan civilizations. The Bible, however, missed it by quite a bit. A verse in the Old Testament, 1 Kings 7:23, implies the value $\pi = 3$. Now perhaps that was an excusable mistake but the following one was not. In 1897, a bill was introduced in the Indiana state legislature which required that $\pi = 3$. The bill passed the House of Representatives by a 63 to 0 vote. Fortunately, a mathematics professor from Purdue University arrived on the scene in the nick of time. Thanks to his intervention, the $\pi = 3$ bill was withdrawn from deliberation by the Indiana Senate and, to date, has not been considered further.

It turns out that π is an *irrational* number, that is, it cannot be expressed as a ratio of two integers, like 22/7. In addition, it is a *transcendental* number, which means that it is not the root of an algebraic equation with integer coefficients, such as $x^2 - 7x + 12 = 0$. The consequence of this transcendental property of π is that the numbers to the right of the decimal point (i.e., 3.14159...) go on forever and ever without any apparent order or pattern.

It has long been a kind of contest or competitive game among mathematicians—and now computer specialists—to enlarge the number of decimal places of π. In the year 1560, it had been established that $\pi = 3.141592$, that is, it was known to an accuracy of six numbers to the right of the decimal point. By the end of the sixteenth century, π had been calculated to thirty decimal places. A summary of the growth of our knowledge of the number of known decimal places of π prior to the twentieth century is presented in table 7.1.

TABLE 7.1

Number (N) of known decimal places of π before the twentieth century

Year	N
1560	6
1600	30
1621	35
1700	71
1706	100
1717	127
1800	140
1824	208
1847	248
1853	440
1855	500
1874	707

With the invention and very fast development of electronic computers during the twentieth century, the number of known decimal places of π has increased rapidly and enormously. In 1947, π was known to 808 places. By 1957, the number had increased to over 10,000 and by 1967 to 500,000. Twenty years later, in 1987, the number had grown to 25 million. In 1997, Professor Kanada and his colleagues at the University of Tokyo computed π to an incredible 51.5 billion decimal places.

One wonders why they want to have all this information about the number of known decimal places of π. Well, for mathematicians involved in number theory, the extensive list of decimal places provides very useful information concerning patterns, distributions, randomness, and other properties and features of number sequences.

Over the years, many analytical methods and mathematical equations have been developed and utilized to calculate π. For example, in the past the following expression has been employed to determine the value of π:

$$\frac{\pi}{4} = 1 - \frac{1}{3} + \frac{1}{5} - \frac{1}{7} + \cdots.$$

(7.1)

This kind of equation is called an infinite series. Another well-known expression utilized for the computation of π is

$$\frac{\pi^2}{6} = 1 + \frac{1}{2^2} + \frac{1}{3^2} + \frac{1}{4^2} + \cdots. \tag{7.2}$$

You might want to try calculating π from equations (7.1) and (7.2). You will quickly discover that they are very slow in producing an answer. Indeed, a good many terms must be employed to obtain even a rough estimate of π. An equation that is much more suitable is the one employed by the noted German mathematician Carl Friedrich Gauss (1777–1855):

$$\frac{\pi}{4} = \arctan(1/2) + \arctan(1/5) + \arctan(1/8). \tag{7.3}$$

With regard to the use of infinite series for calculating the value of π, a remarkable advance was made in 1995 when the following expression was given by Bailey et al. (1997):

$$\pi = \sum_{k=0}^{\infty} \frac{1}{16^k} \left(\frac{4}{8k+1} - \frac{2}{8k+4} - \frac{1}{8k+5} - \frac{1}{8k+6} \right).$$
$$\tag{7.4}$$

Although this equation is only slightly more complicated than the preceding expressions, it yields the value of π much more quickly. You might want to convince yourself that π is correctly computed to six decimal places by using simply the terms corresponding to $k = 0, 1, 2, 3$ in equation (7.4).

A charming little book by Beckmann (1977) gives a brief history of π and descriptions of numerous interesting things about it. In addition, the book contains various mnemonic devices for π. As we know, such devices help us remember things. For example, the mnemonic BASMOQ PN^{3}T^2 gives the provinces and territories of Canada. There are a great many mnemonics for π. The following one helps us remember its numerical value to fourteen decimal places. The number of letters in each word gives the respective number in the sequence (i.e., 3.14159 26535 8979).

How I want a drink, alcoholic of course, after the heavy lectures involving quantum mechanics.

If you prefer a shorter mnemonic device for π, here is one that will give you seven decimal places (3.1415926):

May I have a large container of coffee?

A substantial contribution was made to mathematics literature with the publication of *Pi*: *A Source Book* by Berggren, Borwein, and Borwein (1997). This voluminous work presents the history of this important number over the past 4,000 years. Included in its contents are seventy representative documents on the subject. Most of the contents, of course, deal seriously with the mathematical and computational aspects. For example, the contributions of the Indian mathematical genius Srinivasa Ramanujan (1887–1920) are included in the book.

A good many of the documents in the work deal with strictly historical studies. Thus, considerable attention is given to studies carried out long ago in Egypt, Greece, India, China, and medieval Islam. And finally, a number of presentations are somewhat whimsical or even amusing selections. These include a 402-word mnemonic for π, constructed in the format of a circle, and a display of the numerous documents presented to the Indiana legislature in 1897 to decree the legal value of π.

The Second Most Famous Number: *e*

Although nearly everyone knows about the number π, its nearest rival in fame and importance, the number e, is virtually unknown outside of mathematics, science, and engineering. The main reason for this is that e, which has the approximate value $e = 2.71828$, is not really encountered or utilized—except in natural logarithms—until we get involved in calculus and other areas of more advanced mathematics. In these subjects, e is an extremely important numerical constant.

The Swiss mathematician Leonhard Euler (1707–1783) was one of the most prolific in all of history. Among a great many other major contributions, he was the one who assigned the symbol e to this famous number and proved the relationship

$$\lim_{n \to \infty} \left(1 + \frac{1}{n}\right)^n = e. \qquad (7.5)$$

This expression says that as the whole number n increases in magnitude, the value of the quantity in parentheses—call it S if you like—approaches the number $e = 2.81828$. For example, if $n = 10$, then $S = 2.59374$; if $n = 100$, $S = 2.70481$; if $n = 1,000$, $S = 2.71692$; and so on to $n = \infty$.

Another great mathematician, England's Isaac Newton (1643–1727), showed that

$$e = 1 + \frac{1}{1!} + \frac{1}{2!} + \frac{1}{3!} + \frac{1}{4!} + \cdots. \qquad (7.6)$$

This expression is another example of an infinite series. From either equation (7.5) or equation (7.6), we can easily compute the value of e. As is true of π, the quantity e is an irrational and also transcendental number.

The topic of logarithms was mentioned above. You probably remember logarithms from your course in elementary algebra though you may not recall that you probably dealt only with so-called "common" logarithms. This is the system in which the number 10 is used as the *base* of the logarithm—probably because humans have always had 10 fingers.

Now any number can serve as the base of an arithmetic system including operations involving logarithms. In information theory and computer science, the number 2 ("binary") is generally used although sometimes the numbers 8 ("octal") and 16 ("hexadecimal") are employed. In contrast to the use of the number 10 as the base for "common" logarithms, the number e is used as the base for so-called "natural" logarithms. Fine. But why is such a strange number used for this purpose?

A complete answer to this question involves the limiting value property of e expressed by equation (7.5). To be brief, it is simply pointed out that if we use the numerical value $e = 2.71828\ldots$, then the calculus operations called differentiation and integration are greatly simplified.

For example, suppose we have the simple equation $y = c^x$, where c is a positive constant. If we set $c = e$, then clearly $y = e^x$. This is called the exponential function. Here comes the calculus. The derivative of this equation is $dy/dx = e^x$. In other words, the derivative of the exponential function is equal to the function itself. By the same token, the integral of e^x takes on the same form, that is, $\int y\, dx = e^x$. Thus, by using this particular numerical value for e, the function e^x, its derivative, and its integral are all equal. For calculus operations, this represents an enormous simplification.

Furthermore, with $y = e^x$, then taking logarithms, we have $x = \log_e y$, where the subscript e means that the base of the logarithm is e. This is a "natural" logarithm. Incidentally, $\log_e y$ is sometimes written $\ln y$ to avoid confusion with $\log_{10} y$, the so-called common logarithm. Your calculator probably has keys for both types of logarithms.

An Example: Earning Interest on Your Savings Account

We take a quick look at a topic in which all of us are interested: how much money can you earn on your savings account? To answer the question, we utilize equation (7.5) but change it slightly to the form

$$P = P_0\left(1 + \frac{r}{n}\right)^n, \tag{7.7}$$

in which $P_0 = \$1{,}000$ is the amount of money you have in your savings account at the beginning of the year, that is, the original principal; $r = 6\% = 0.06$ is the annual interest rate paid by your bank; n is the number of "compounding periods" during the year; and P is the amount of money in your account at the end of the year, i.e., 12 months later. The difference between P and P_0 is the amount of interest you earned during the year.

Now if the bank compounds your interest earnings only once a year, then $n = 1$. Accordingly, from equation (7.7), at the end of

TABLE 7.2

Interest earnings on your savings account
Original principal P_0 = \$1,000; interest rate r = 6%.

Compounding frequency	n	Year-end principal
Annually	1	\$1,060.00
Semiannually	2	1,060.90
Quarterly	4	1,061.36
Monthly	12	1,061.68
Daily	365	1,061.83
Instantly	∞	1,061.84

the year you will have $P = \$1,000(1 + 0.06/1)^1 = \$1,060$. This is called simple interest. Alternatively, suppose the bank compounds semiannually. In this case, $n = 2$ and equation (7.7) becomes $P = \$1,000(1 + 0.06/2)^2 = \$1,060.90$. Next, assume that the bank compounds quarterly. Therefore, $n = 4$; substituting into equation (7.7) gives $P = \$1,000(1 + 0.06/4)^4 = \$1,061.36$, and so on. The results of our calculations are listed in table 7.2.

Observation number 1: Suppose that your bank advertises and applies "daily compounding" of interest on your savings account. Then, after 12 months, according to the table, your principal is \$1,061.83 and you have earned $P - P_0 = \$61.83$. Dividing this by the original principal and converting to a percentage gives $r_* = 6.183\%$. This is called the *yield*.

Observation number 2: If there is instantaneous compounding, equation (7.7) becomes $P = P_0 e^r$, which gives $P = \$1,061.84$. Clearly, as far as your interest earning is concerned, it makes essentially no difference whether your bank compounds your savings account daily, hourly, or instantaneously.

Another Example: Geometrical Interpretation of the Number e

Everyone knows that π has a very simple geometrical interpretation. It gives the circumference of a circle in terms of its diameter, $C = \pi D$. Alternatively, it expresses the area of a circle

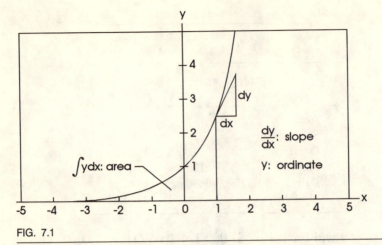

FIG. 7.1

A plot of the exponential function showing the geometrical interpretations of the ordinate, slope, and area under the curve.

in terms of its radius, $A = \pi R^2$. So a logical question to ask is, what is the simplest geometrical interpretation you can devise involving that other very important number, e?

To get things started, here is one idea. As we know, the equation of the exponential function is $y = e^x$; this expression is displayed graphically in figure 7.1. Now since $y = e^x$, when $x = 1$, $y = e$. So, as shown in figure 7.1, e is the value of the *ordinate*, y, when the *abscissa*, $x = 1$. Further, as pointed out earlier, if $y = e^x$, then the first derivative is $dy/dx = e^x$. It turns out that the first derivative of a function represents the *slope* of the function. Accordingly, at $x = 1$, $dy/dx = e$ is the slope of the curve, as shown in the figure. Well, here are two (not terribly exciting) geometrical interpretations of e: the ordinate and the slope of the exponential function at $x = 1$.

Going further, if $y = e^x$ then an integration gives the result $\int y\,dx = e^x$ if the lower limit of the integral is $x = -\infty$. The geometrical interpretation of this result is that the total area under the curve between $x = -\infty$ and $x = x$ is $A = e^x$. If we again select $x = 1$, then $A = e$ is the *area under the curve* between $x = -\infty$ and $x = 1$. This is another geometrical interpretation of e.

You can do much better than this and here is your chance to prove it.

ANNOUNCEMENT OF A GREAT CONTEST INVOLVING
THE NUMBER e

   Part 1. Devise some kind of relationship that gives a simple geometrical interpretation of e.

   Part 2. Manufacture a mnemonic device for e along the lines given earlier for π. For your information, $e = 2.71828\ 18284\ 59045$.

An interesting reference, somewhat analogous to the one by Beckmann (1971) concerning π, is the book by Maor (1994), which deals with our other famous number, e.

Three Other Famous Numbers

We have taken quick looks at the two most famous numbers in mathematics: π and e. There are numerous others; if you are interested in delving further into the matter, a good place to start is the little book by Wells (1986). For the present, we mention three other important numbers.

Golden Ratio, $\phi = 1.61803$

This number defines the ratio of the length L and width H of a rectangle that allegedly gives the most esthetically attractive appearance. Its precise value is $L/H = \frac{1}{2}(1 + \sqrt{5})$. We examine this number in detail in chapter 9, "Great Number Sequences: Prime, Fibonacci, and Hailstone."

Euler's Constant, $\gamma = 0.57721$

This important number, devised by Leonhard Euler around 1750, is defined by the equation $\gamma = (1 + 1/2 + 1/3 + \cdots + 1/n - \log_e n)$ as n becomes infinite. It makes its appearance in many problems in mathematics and statistics.

Feigenbaum Number, $\delta = 4.66920$

This is the newest of the important numbers. It was discovered in 1978 by the American mathematician Mitchell Feigenbaum, in his early studies of chaos theory. This number, δ, is the ratio of the spacings of successive intervals of period doubling in a process leading to chaos. An easily understood description of the nature of δ is presented by Addison (1997).

Some References for More Information

In this brief look at some of the famous numbers of mathematics, we have run across the names of Archimedes, Euler, Gauss, and Newton. These are probably the four greatest mathematicians of all time. If you would like to learn more about them, you will find the book by Muir (1996) to be helpful. It presents interesting biographies of these four notable mathematicians and quite a few others. In addition, the comprehensive work of Boyer (1991) presents a great deal of information about the matter.

Annual Celebration Days for π and Perhaps e

In the preceding sections, we have examined two famous numbers, π and e, and looked briefly at three others, ϕ, γ, and δ. Noteworthy is the fact that of these five important numbers, only π can be expressed as a date if we require that three significant figures be utilized. That is, $\pi = 3.14$ which, of course, is March 14.

It is fitting, therefore, that this most important number of all should be celebrated each year. Accordingly, shall we declare March 14 to be π-Day? If this meets with success, we could, later on, celebrate e-Day on February 7 and perhaps ϕ-Day on January 6. Suggested first steps: Contact the White House and the greeting card manufacturers.

8

Another Look at Some Famous Numbers

In the preceding chapter we examined several of the most important numerical constants appearing in mathematics. The best known of these "famous numbers" are, of course, π (the ratio of the circumference and diameter of a circle) and e (the base of natural logarithms). We also looked briefly at three other important numbers: ϕ (the golden ratio), γ (Euler's constant), and δ (the Feigenbaum number).

In this chapter, the subject of famous numbers is continued but aimed in a somewhat different direction. We begin with what are called real numbers, imaginary numbers, and complex numbers. First, some rules are given for procedures such as the addition of complex numbers—so-called vector addition. Then a very interesting equation, originally given by Leonhard Euler, is introduced to generate some remarkable relationships involving real and imaginary numbers, including the one presented in the opening paragraph of the previous chapter. Using the methods of complex number addition, and calling on some things we know about π and e, we construct a polygonal spiral which we then use wickedly to strangle a negative number.

Finally, suppose you had to make a quick decision regarding a gift from your wealthy uncle. Which would you rather receive from him: 4^3 dollars or 3^4 dollars? Expressing this question in general terms and assuming that a is larger than b, which is the greater quantity: a^b or b^a?

Numbers: Real, Imaginary, and Complex

We start with a very simple algebraic equation: $x^2 - 1 = 0$. What is its solution? Easy: We move the 1 to the other side of the equation to get $x^2 = +1$ and then take the square root of both sides. This yields $x = \sqrt{+1} = 1$. However, don't forget that it also yields $x = -1$, because $(-1) \times (-1) = +1$. So the solution to this little equation is $x = \pm\sqrt{+1} = \pm 1$.

Fair enough. Now suppose we start with another very simple equation: $x^2 + 1 = 0$. Moving the 1 to the other side gives the expression $x^2 = -1$, and taking the square root yields $x = \pm\sqrt{-1}$. What in the world does $\sqrt{-1}$ mean? Well, this is called an "imaginary" number; it may or may not be a good name but we're stuck with it. Very simply, an imaginary number is the square root of a negative number. For hundreds of years, mathematicians did not know how to handle imaginary numbers, though nowadays they present no particular difficulties. Indeed, in many areas of mathematics, imaginary numbers are extremely useful quantities. The letter i has been adopted as the symbol for imaginary numbers. That is, $i = \sqrt{-1}$. Furthermore, an entire methodology has been developed to handle arithmetical and algebraic manipulations involving imaginary numbers.

For example, consider the following equation: $z_1 = +5 - 2i$. In this expression, the $+5$ is called a *real* number. On the other hand, the $-2i$ is termed an *imaginary* number because it contains the quantity i. The indicated sum, z_1, is called a *complex* number.

Now consider another equation: $z_2 = -3 + 4i$. Suppose we want to add z_1 and z_2. To do this, we simply add the real numbers of the two equations and then add the imaginary numbers of the two equations. In our example, this operation yields the equation $z_1 + z_2 = +2 + 2i$. All this is shown graphically in figure 8.1, in which the real numbers are plotted along the x-axis and the imaginary numbers are plotted along the y-axis. The

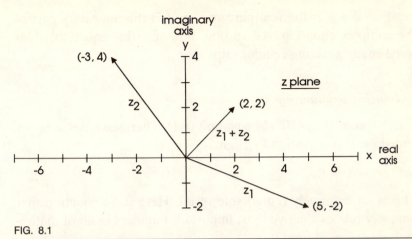

FIG. 8.1

An example of vector addition. $z_1 = 5 - 2i$, $z_2 = -3 + 4i$, so $z_1 + z_2 = 2 + 2i$. This plot is an Argand diagram.

$z = x + iy$ plane is called the complex plane. We sometimes refer to z_1 and z_2 as *vectors*. We have just carried out a simple example of *vector addition*. We can also perform vector subtractions, multiplications, and divisions. The display shown in figure 8.1 is sometimes called an Argand diagram.

For use in an example we shall look at in a moment, we summarize the immediately preceding analysis. In a rectangular (x, y) coordinate system we plot real numbers in the plus or minus x-direction depending on whether they are positive or negative quantities. In the same way, we plot imaginary numbers in the plus or minus y-direction depending on whether they are positive or negative quantities. What could be simpler? Incidentally, a suggested reference for an elementary presentation of various topics concerning imaginary and complex numbers is Gardner (1992).

Some Amazing Mathematical Relationships

In 1746, that incredibly prolific mathematical genius named Leonhard Euler presented the following identity:

$$e^{i\theta} = \cos \theta + i \sin \theta, \tag{8.1}$$

in which cos θ is the real part and sin θ is the imaginary part of the complex equation. As we shall now see, this equation yields some really amazing relationships.

Amazing Relationship 1

Let $\theta = \pi$ (i.e., 180°) in equation (8.1). Then since cos $\pi = -1$ and sin $\pi = 0$, equation (8.1) gives

$$e^{i\pi} + 1 = 0. \tag{8.2}$$

This is an extremely remarkable result. Here is an equation that uniquely relates the five most important numbers in all of mathematics: e, i, π, 1, and 0.

Amazing Relationship 2

Let $\theta = \pi/2$ (i.e., 90°) in equation (8.1). Then since $\cos(\pi/2) = 0$ and $\sin(\pi/2) = 1$, we get $e^{i\pi/2} = i$. Multiplying the exponents of both sides of this equation by i yields $e^{-\pi/2} = i^i$. Since $e^{-\pi/2} = 0.2079$, we obtain the very remarkable result that $i^i = 0.2079$. This relationship says that i raised to the ith power is equal to a real number. Totally crazy! How can this possibly be? This is much too weird to even think about.

How to Strangle a Negative Number

We are now going to strangle the number -1, and here is how we are going to do it. First, the following is a generalization of the infinite series given by equation (7.6):

$$e^{\theta} = 1 + \frac{\theta}{1!} + \frac{\theta^2}{2!} + \frac{\theta^3}{3!} + \cdots . \tag{8.3}$$

Next, from equation (8.2) we get the relationship $e^{i\pi} = -1$. So if we substitute $\theta = i\pi$ into equation (8.3) we obtain

$$e^{i\pi} = 1 + \frac{(i\pi)}{1!} + \frac{(i\pi)^2}{2!} + \frac{(i\pi)^3}{3!} + \cdots = -1. \tag{8.4}$$

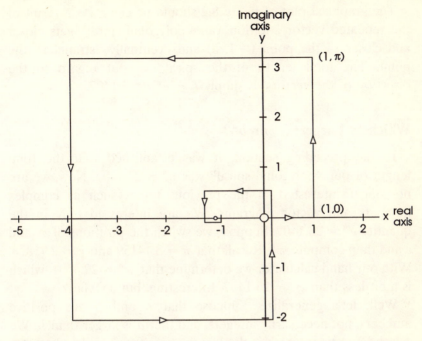

FIG. 8.2

The polygonal spiral created by vector addition of the infinite-series form of $e^{i\pi}$. Note that the spiral converges on the point $(-1, 0)$.

Now remember that $i = \sqrt{-1}$. Therefore $i^2 = -1$, $i^3 = -i$, $i^4 = +1$, $i^5 = +i$, $i^6 = -1$, and so on. Substituting these values into equation (8.4) yields

$$1 + i\pi - \frac{\pi^2}{2} - i\frac{\pi^3}{6} + \frac{\pi^4}{24} + i\frac{\pi^5}{120} - \cdots = -1. \qquad (8.5)$$

We now prepare a graphical display of this equation, following the rules we presented and utilized in constructing figure 8.1, in which we simply added z_1 and z_2. The results of this exercise in vector addition are shown in figure 8.2. The angles at the corners of the resulting polygon are all 90°. The lengths of the sides are 1, π, $\pi^2/2$, $\pi^3/6$, $\pi^4/24$, and so on. The coordinates of the first two corners are indicated in the figure; the others are easily calculated.

The graphical plot of figure 8.2 should be clear. As a result of the repeated vector addition, our "polygonal spiral" gets closer and closer to the point $(-1, 0)$ and eventually "strangles" the point. The total length of the spiral, without regard to the direction of the vectors, is simply $L = e^\pi = 23.1407$.

Which Is Larger: a^b or b^a?

In the preceding section, it was established that the total length of our "polygonal spiral" was $e^\pi = 23.1407$. Now, we are no longer interested in the previous topics such as complex numbers and vector diagrams. We are interested only in the quantity $e^\pi = 23.1407$. Suppose we switch the positions of π and e and then compute π^e. Recall that $\pi = 3.14159$ and $e = 2.71828$. With our hand calculator we determine that $\pi^e = 22.4591$, which is a bit less than $e^\pi = 23.1407$. Interesting, but so what?

Well, let's generalize. Suppose that a and b are positive numbers, not necessarily integers, and that a is larger than b. We ask the question: which is the larger quantity, a^b or b^a, that is, a to the bth power or b to the ath power? Incidentally, this problem was suggested by two somewhat similar problems that appear in Dörrie (1965) and Dunn (1980).

To start with, here are three numerical examples:

1. If $a = 3$ and $b = 2$, then $a^b = 3^2 = 9$ and $b^a = 2^3 = 8$. In this case, $a^b > b^a$.

2. If $a = 4$ and $b = 2$, then $a^b = 4^2 = 16$ and $b^a = 2^4 = 16$. In this case, $a^b = b^a$.

3. If $a = 5$ and $b = 2$, then $a^b = 5^2 = 25$ and $b^a = 2^5 = 32$. In this case, $a^b < b^a$.

We note that modest changes in the magnitude of a, with the value of b held constant in these three examples, entirely alter the relative magnitudes of a^b and b^a. This seems strange.

It will be very helpful in our analysis if we know when the two quantities, a^b and b^a, are equal. That is,

$$a^b = b^a. \tag{8.6}$$

Unfortunately, it is impossible to obtain a solution to this equation in a form that expresses a in terms of b or vice versa. Consequently, we must settle for a *parametric* solution. Taking the logarithms of both sides of equation (8.6) gives

$$b \log_e a = a \log_e b. \tag{8.7}$$

We let $a = kb$, where k is a number larger than one (since $a > b$); k is called the parameter. Substituting this relationship into equation (8.7) and carrying out some algebra, we obtain the expressions

$$a = k^{k/(k-1)}; \; b = k^{1/(k-1)}; \; k > 1. \tag{8.8}$$

Assigning various numerical values to the parameter k in these expressions gives the magnitudes of a and b that satisfy equation (8.6).

An interesting question is, what are the values of a and b when k approaches one? We note that the relationships of equation (8.8) are indeterminate for $k = 1$. Never mind. In these two expressions, we make the substitution $n = 1/(k - 1)$. This yields

$$a = \left(1 + \frac{1}{n}\right)^{n+1}; \; b = \left(1 + \frac{1}{n}\right)^{n}. \tag{8.9}$$

It is clear that if $k = 1$ then $n = \infty$. Utilizing equation (7.5), it is easily established that in this limiting ($k = 1$) case we have the values $a = e$ and $b = e$.

The results of our analysis are displayed in figure 8.3. For your information, table 8.1 lists some coordinates for the curve $a^b = b^a$ shown in the figure. Two interesting observations: (1) The values $a = 4$, $b = 2$ are the only *integers* for which $a^b = b^a$. (2) The values $a = e$, $b = e$, as the end point ($k = 1$) coordinates on the $a^b = b^a$ curve, represent an unexpected appearance of our very remarkable number e.

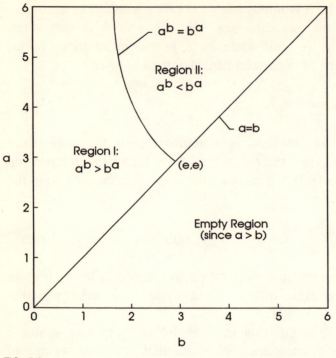

FIG. 8.3

Graphical display of the regions in which a^b is greater than, equal to, and less than b^a.

TABLE 8.1

Coordinates for the curve $a^b = b^a$ shown in figure 8.3

k	a	b
1.0		
1.1	2.853	2.594
1.2	2.986	2.488
1.6	3.502	2.189
2.0	4	2
3.0	5.196	1.732
4.0	6.349	1.587
6.0	8.586	1.431
10.0	12.915	1.292
∞	∞	1

The main feature of the display of figure 8.3 is the identification of region I in which $a^b > b^a$ and region II where $a^b < b^a$. You might want to confirm these results by computation with some selected values for a and b, including $a = \pi$ and $b = e$. Also, you will undoubtedly want to show that the slope of the $a^b = b^a$ curve at the point (e, e) is $(da/db) = -1$, i.e., it is perpendicular to the line $a = b$.

9

Great Number Sequences: Prime, Fibonacci, and Hailstone

What Are Number Sequences?

The answer to this question is that number sequences are simply lists of numbers appearing in a particular order. For example, the sequence 0, 1, 2, 3, 4, 5, 6, 7, 8, 9 is the list of the natural numbers or integers. The sequence 1, 3, 5, 7, 9, 11 is the list of the odd integers and the sequence 1, 4, 9, 16, 25, 36 is the list of the squares of the integers.

Another example: The sequence 1, 2, 4, 8, 16, 32, 64 is the list of numbers generated by doubling successive numbers. Sometimes, as in this case, a mathematical formula can produce the sequence of numbers. In this example, the formula is $N = 2^n$, where N is the magnitude of the nth number of the sequence. This particular sequence is the one that describes geometric growth; it is closely related to so-called exponential growth. This sequence is quite famous. In a quantitative way, it reflects the gloomy prediction of Thomas Malthus, the eighteenth-century English clergyman-economist, regarding the explosive growth of human population.

Here is an interesting sequence: 6, 28, 496, 8,128, and so on. This is the sequence of so-called *perfect numbers*. A perfect number is a number composed of the sum of all its divisors except the number itself. For example, it is clear that 6 and 28 are perfect numbers because $6 = 1 + 2 + 3$ and $28 = 1 + 2 + 4 + 7 + 14$.

These numbers were known and studied by the Greeks more than two thousand years ago. An interesting fact is that the next perfect number in the sequence, following the 8,128, is 33,550,336. After that come 8,589,869,056 and then 137,438,691,328. At present (1999), we know the values of thirty-six perfect numbers. In recent years, the rapid development of computers has accelerated our search for these numbers.

There are a great many sequences in mathematics; some are important or interesting, others are trivial or silly. Over the years, a very large number of sequences have been discovered or devised. A collection of over 2,300 from all branches of science and mathematics has been nicely assembled by Sloane (1973). On the following pages, we shall investigate a few.

Mostly about Prime Numbers

Suppose we have the following sequence of positive numbers: 1, 2, 3, 5, 7, 11, 13, 17, 19, 23, 29, 31, 37, and so on. Does this sequence represent anything in particular?

It certainly does. It is the sequence of so-called *prime* numbers. That is, it is a list of the numbers that are exactly divisible only by 1 and the number itself. For example, 13 is a prime number because it can be exactly divided only by 1 and 13. On the other hand, 12 is not a prime number; in addition to 1 and 12, it can also be divided by 2, 3, 4, and 6.

The natural numbers or integers are 1, 2, 3, 4, 5, 6, 7, 8, and so on. If a natural number is neither 1 nor a prime it is called a *composite* number. Following the 37, the above sequence of prime numbers continues: 41, 43, 47, 53, 59, 61, 67, and so on. Altogether, there are 168 prime numbers less than 1,000 and, with the exception of 2, they are all, of course, odd numbers. By the way, mathematicians usually do not consider 1 to be a prime number.

It turns out that the sequence of prime numbers goes on forever. The famous Greek geometer Euclid was the first to prove this; the year, around 300 B.C. Ever since then, mathematicians have been trying to discover or develop a method or an equation to determine whether or not a particular number is a prime number. So far, they have not been successful.

Long ago, numerous formulas were devised that generated prime numbers. A good example is the expression $N = n^2 - 79n + 1601$. This formula produces prime numbers for all values of n up to and including $n = 79$. You might want to check a few. However, when $n = 80$, the formula fails because $N = 1681 = 41 \times 41$. Thus, 1681 is a composite number. Other early expressions for generating prime numbers were $N = n^2 - n + 41$ and $N = n^2 + n + 17$. With the benefit of hindsight, it now seems obvious that these polynomial-type expressions are entirely inadequate in producing a lengthy list of only prime numbers.

The noted French mathematician Pierre de Fermat (1601–1665) also searched for a formula that would yield only prime numbers. He invented the following equation, which produces what we now call the Fermat numbers:

$$F_n = 2^{2^n} + 1, \ n = 1, 2, 3 \ldots . \tag{9.1}$$

This expression generates the numbers $F_0 = 3$, $F_1 = 5$, $F_2 = 17$, $F_3 = 257$, and $F_4 = 65,537$. All of these numbers are prime. However, F_5—which is a ten-digit number—is not prime and neither are the Fermat numbers corresponding to $n = 6, 7, 8, 9, 11$, and many more numbers calculated with larger values of n. So Fermat was mistaken about equation (9.1) producing only prime numbers. Indeed, a present-day question in number theory is, are *any* Fermat numbers, larger than $n = 4$, prime numbers? So far, the question has not been answered.

Mathematicians have long since abandoned the search for formulas that generate prime numbers; there are a great many more interesting subjects in number theory for them to investigate. The following is an important example.

How Are the Prime Numbers Distributed?

It is well known that the sequence of prime numbers goes on forever. However, as the numbers become larger and larger, their frequency decreases, that is, the gaps between successive primes increase. This feature provided the historical basis for the statement of the fundamental theorem of prime numbers. This theorem, for which there are now numerous proofs, postulates that as n increases to very large values, the number of primes not exceeding n is given by the expression

$$P(n) = \frac{n}{\log_e n}.$$

(9.2)

The problem attracted the attention of many of the leading mathematicians of the past. The German mathematician Carl Friedrich Gauss (1777–1855), one of the greatest of all time, examined the problem of prime number distribution in 1792 when he was only fifteen years of age.

Table 9.1 may help illustrate our point concerning the distribution of primes. In the table, column 2 indicates that there are 168 prime numbers less than 1,000. Column 3 says that the number of primes, computed from equation (9.2), is 145, which is somewhat

TABLE 9.1

Distribution of prime numbers

(1) Magnitude of n	(2) Primes less than n	(3) $P(n) = \dfrac{n}{\log_e n}$	(4) $\dfrac{\text{Column 3}}{\text{Column 2}}$	(5) $P(n) = Li(n)$
1,000	168	145	0.863	175
10,000	1,229	1,086	0.883	1,245
100,000	9,592	8,686	0.906	9,630
1,000,000	78,498	72,382	0.922	78,628
10,000,000	664,579	620,421	0.934	664,918
100,000,000	5,761,455	5,428,675	0.942	5,762,209

Source: Lines (1986).

less than the actual number. As shown in column 4, the ratio of column 3 to column 2 is 0.863.

Now observe that as the number n increases, the ratio of the two quantities of prime numbers, shown in column 4 of the table, gradually increases and appears to approach the limiting value 1.0. Indeed, if $n = 1$ billion the ratio is 0.949 and if $n = 10$ billion the ratio is 0.954. This is precisely the result postulated by the prime number theorem.

The entire matter can be carried a significant step further. Studies have shown that an even more accurate answer to the prime number theorem is given by the expression

$$P(n) = \text{Li}(n) = \int_2^n \frac{dx}{\log_e x}, \tag{9.3}$$

where $\text{Li}(n)$ is the so-called logarithmic integral. The indicated integral is tabulated in numerous mathematical handbooks.

For the sake of completeness, the number of primes determined by equation (9.3) is listed in column 5 of table 9.1. If these numbers are compared with those shown in column 2 of the table, we note very close agreement. Thus, the accuracy of equation (9.3) in predicting the distribution of prime numbers is quite remarkable.

What Is the Largest Prime Number?

Of course, one of the greatest challenges to mathematicians is to discover ever-larger prime numbers. Over the years, the magnitude of the largest known prime number has steadily increased; clearly, the very rapid advances in computer technology have greatly assisted these efforts. Currently (1999), the largest known prime number is $2^{3,021,377} - 1$. This number is so enormous that it requires 909,526 digits to express it.

Not surprisingly, some of the greatest mathematicians of all time have worked on the theory of prime numbers. If you are interested in the historical aspects of the subject, Bell (1956) and Boyer (1991) are good places to start. Elementary presentations of topics on prime numbers are given by Beiler (1964), Lines (1986), and Ogilvy and Anderson (1988). A more advanced coverage is presented in the very readable book by Ribenboim (1995), which includes a lengthy list of references.

Concerning Fibonacci Numbers

One of the most remarkable sequences of numbers in mathematics is the Fibonacci sequence or simply the Fibonacci numbers.

The story all began about eight hundred years ago when an Italian mathematician named Leonardo of Pisa (1170–1250)—better known as Fibonacci—wrote a book entitled *Liber Abaci* (first published in 1202; revised in 1228). He was born in Pisa, Italy, at about the time construction was begun on what is now called the Leaning Tower of Pisa (which was completed around 1300). Incidentally, he lived long before Galileo Galilei (1564–1642) carried out his famous weight-dropping experiments from the top of the tower.

During much of his early life, Fibonacci lived in North Africa and studied algebra and geometry under Arabic mathematics teachers. As a result, he learned a great deal about the Arabic number system and decimal notation concepts; he included these topics in his *Liber Abaci*. Thank goodness! Without question, he has our eternal gratitude for hastening the demise of the ghastly Roman numeral system. (Try dividing CCCXXI by XLIX.)

In any event, Fibonacci was undoubtedly the greatest European mathematician of the Middle Ages. Among a great many other things, he presented a mathematical problem in his *Liber Abaci* which is still used as a popular way to introduce the subject of Fibonacci numbers. This is his famous rabbit problem, as presented by Vajda (1989):

A pair of newly born rabbits is placed in a confined enclosure. This pair, and every later pair, produces one new pair every month, starting in their second month of age. How many pairs will there be after one, two, three, . . . , months?

TABLE 9.2

Fibonacci's rabbits

End of month number	Number of pairs of rabbits
1	1
2	2
3	3
4	5
5	8
6	13
7	21
8	34
9	55
10	89

Source: Vajda (1989).

The answer to the problem is displayed in table 9.2. From the table, we note that the sequence representing the number of pairs of rabbits is 1, 2, 3, 5, 8, 13, 21, 34, and so on. These are the famous Fibonacci numbers. Without doubt, you have already discovered how the sequence increases: each number is the sum of the two previous numbers. So the recurrence relationship is

$$F_{n+1} = F_n + F_{n-1}. \tag{9.4}$$

The next few numbers are 144, 233, 377, 610, and so on. Now here is an important point. If you divide each number in the sequence by the preceding number, you get closer and closer to the quantity 1.618034. This is a very interesting result; we shall come back to it shortly.

The diversity of places where the Fibonacci numbers make appearances is absolutely incredible. In one form or another they show up not only in numerous topics of mathematics but also in biology, botany, music, art, and architecture. At the elementary level, suggested references are Huntley (1970), Lines (1990), and Dunlap (1997); more advanced coverage is given by Vajda (1989). It is noteworthy that in 1963, a Fibonacci Association was cre-

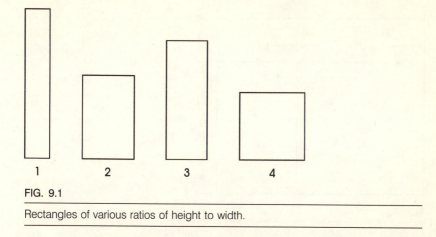

1   2   3   4

FIG. 9.1

Rectangles of various ratios of height to width.

ated, which began publication of the *Fibonacci Quarterly*. Over the years, hundreds of articles have been published in the *Quarterly* dealing with a great many subjects involving Fibonacci numbers.

Closely related to the Fibonacci sequence is one called the Lucas sequence. As before, each term of this sequence is the sum of the previous two terms. However, instead of starting with the numbers (1, 2) as in the Fibonacci sequence, the Lucas sequence begins with (1, 3). This yields 1, 3, 4, 7, 11, and so on.

Fibonacci Numbers and the Golden Section

In figure 9.1, four rectangles are shown with height-to-width ratios ranging from the tall slender rectangle on the left (1) to the square on the right (4). Which one of the four shapes do you like the best? That is, which, if any, do you think is the most esthetically appealing?

Well, different people like different things, but psychologists have found that most people "like" rectangle 2 best, because it is neither too slender nor too square; it "seems about right." In any event, rectangle 2 of figure 9.1 is rotated 90° and dimension symbols added in figure 9.2.

The ancient Greeks, especially their architects, were probably the first to believe that there was a ratio between the length and

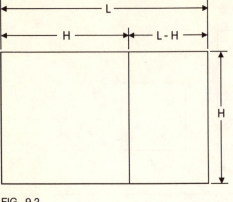

FIG. 9.2

Definition sketch for the golden rectangle or golden section.

the height of a rectangle that gave the most pleasing artistic proportion. Historians suspect that the famous Parthenon in Athens was designed and built with this ratio in mind.

With reference to figure 9.2, our problem commences by simply constructing the pleasing ratio

$$\frac{L}{H} = \frac{H}{L-H}. \tag{9.5}$$

From here on, the problem is one of algebra. The above equation yields the expression $L^2 - HL - H^2 = 0$. If you remember how to solve quadratic equations like this one, you will easily obtain the answer,

$$\frac{L}{H} = \frac{1}{2}(1 + \sqrt{5}). \tag{9.6}$$

So we have determined the length-height ratio of a rectangle that gives the most pleasing appearance, at least in the opinion of the early architects and artists. This ratio, L/H, has been given a special name: the golden ratio or golden section. Furthermore, since mathematicians consider it to be a very important constant, it has also been given a special symbol, ϕ. That is,

$$\phi = \frac{1}{2}(1 + \sqrt{5}).$$ (9.7)

Now, are you ready for an interesting surprise? The numerical value of the golden section is $\phi = 1.618034\ldots$. Does this quantity look familiar? It is the ratio of successive Fibonacci numbers we looked at a few moments ago. Fantastic, right? Why should the breeding habits of a bunch of prolific rabbits have anything in common with ancient Greek architecture? Mathematics is not only beautiful, it can also be very intriguing!

A few more items about ϕ. The early Egyptians may also have been aware of the golden section. In chapter 20, "How to Make Mountains Out of Molehills," we discuss the fact that the main dimensions of the Great Pyramids of Egypt (built during the period 2650 to 2500 B.C.) seem to conform to the geometrical proportions directly related to ϕ.

On quite another topic and as we established in chapter 1 the perimeter P of a regular five-pointed star of radius R is $P = 10\sqrt{(2\phi - 1)/(2\phi + 1)}\,R$. The dimensions of playing cards are not greatly different from the ratio ϕ. Most books seem to have this proportion, and virtually all the flags of the world have length:height ratios close to ϕ:1.

The golden ratio or golden number, ϕ, appears frequently in other quite unexpected places. Quite a few of these are discussed by Huntley (1970); here is one more. Consider a square of unit side length as shown in figure 9.3. Another square of side length x is removed from the first square. The center of gravity of the L-shaped area is at O. What is the value of x?

This problem, proposed by Lord (1995), is solved as follows. We compute the torques (or moments) about axis OO' (or axis OO''). Equating the clockwise torques to the counterclockwise gives

$$(x)(1 - x)\frac{1}{2}x = (1)(1 - x)\frac{1}{2}(1 - x),$$

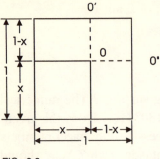

FIG. 9.3

Definition sketch for a balanced area.

which yields

$$x^2 + x - 1 = 0.$$

The solution to this quadratic equation is $x = (1/2)(\sqrt{5} - 1)$. This is the length of the rectangular leg; its width, of course, is $(1 - x)$. A bit of algebra shows that the length-width ratio $x/(1 - x) = (1/2)(1 + \sqrt{5}) = \phi$. Thus, the two rectangular legs of the L-shaped area are golden rectangles: an interesting result. You might want to test this answer by very carefully cutting the L-shaped area from a piece of stiff cardboard and balancing it on a needle at point O.

Finally, as illustrated in figure 9.4, suppose we start with a relatively large golden rectangle of length L and width H. We mark off the corresponding square of side length $L - H$. This leaves another but smaller golden rectangle. We again mark off the corresponding square. This process is repeated until we begin to approach the pole, O. Then a smooth curve is drawn through the corners of each of the squares. It turns out that this curve is what is called an equiangular spiral or logarithmic spiral. It is surely one of the most beautiful curves in all of mathematics. Let us take a closer look at it.

The Golden Section and the Logarithmic Spiral

In the branch of mathematics called analytic geometry, two basic systems are utilized to display and mathematically describe

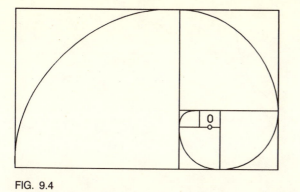
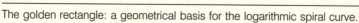

FIG. 9.4

The golden rectangle: a geometrical basis for the logarithmic spiral curve.

plane curves. One is the *rectangular* coordinate system, in which any point P is specified by its rectangular (x, y) coordinates. The other is the *polar* coordinate system, in which a point is identified by its polar (r, θ) coordinates. We shall use polar coordinates in our present analysis.

The pole O and a portion of the curve illustrated in figure 9.4 are shown in figure 9.5. The equation of the logarithmic spiral is

$$r = r_0 e^{\theta \cot \alpha}, \tag{9.8}$$

in which (r, θ) are the polar coordinates, r_0 is the value of r when $\theta = 0$, and α is the angle between the radius vector r and the tangent to the curve at any point P; the symbol $\cot \alpha$ is the cotangent $\alpha = 1/\text{tangent } \alpha$. The fact that this angle is constant all along the curve is the reason that it is sometimes called the equiangular spiral.

It is not difficult to show, using equation (9.8), that the length of the logarithmic spiral is

$$S = (r - r_0)\sec \alpha, \tag{9.9}$$

where $\sec \alpha = \text{secant } \alpha = 1/\text{cosine } \alpha$.

There are numerous interesting features of this beautiful curve. Perhaps the most noteworthy is that the shape of the curve does not change as it increases in size. In nature, this remarkable property is vividly displayed in the Nautilus sea shell. As seen in

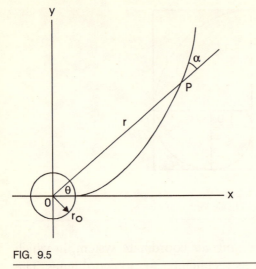

FIG. 9.5

Definition sketch for the logarithmic spiral.

figure 9.6, this shell has the shape of a logarithmic spiral. Other kinds of sea shells, the horns of various types of animals, and growth patterns of sunflowers and other plants all have these spiral shapes. These and related topics are discussed by Huntley (1970) and by Thompson (1961).

We conclude our brief look at logarithmic spirals with a numerical example involving naval warfare tactics.

EXAMPLE: A DESTROYER INTERCEPTS A SUBMARINE.   A destroyer sights an enemy submarine at a distance $L = 4.0$ nautical miles. Knowing that it has been sighted, the submarine quickly dives and heads off at maximum velocity $U_s = 10$ knots, in a certain direction θ. The velocity of the destroyer is $U_d = 30$ knots. Assuming that the submarine's velocity and direction do not change, what should be the destroyer's tactics to assure an interception regardless of the direction the submarine takes?

Our example is illustrated in figure 9.7. The points D and S identify the locations of the destroyer and the submarine at the moment of initial visual contact. Now the submarine, initially at point S, immediately dives after visual contact and proceeds along some constant course,

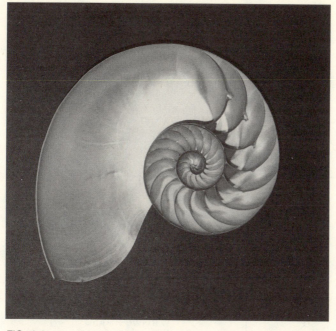

FIG. 9.6

A logarithmic spiral in nature: the Nautilus sea shell.

θ, at a velocity $U_s = 10$ knots. The destroyer, initially at D, should head directly for point S at a velocity $U_d = 30$ knots and maintain that course for 3.0 miles, that is, 6 minutes. (Why? Hint: if the submarine were to select the course $\theta = 0$, (that is, headed directly for the destroyer, interception would occur at point C.)

When the destroyer reaches point C, it should make a sudden radical course change and head initially in the direction $\theta = \alpha = \arccos(U_s / U_d) = 70.5°$. Thereafter, the destroyer should follow the continuously changing course defined by the logarithmic spiral of equation (9.8) in which $r_0 = 1.0$ miles and $\cot \alpha = 0.3536$. The spiral course of the destroyer is shown in figure 9.7.

It should be apparent that the destroyer will intercept the submarine regardless of the direction θ of the latter. You can check your calculations by using equation (9.9) to compute the distances covered by the destroyer. With this information, you can calculate interception times. A typical result: If the submarine were to select a course

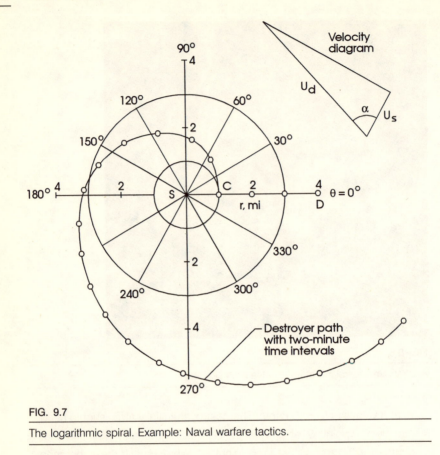

FIG. 9.7

The logarithmic spiral. Example: Naval warfare tactics.

$\theta = 270°$, interception would occur at a distance 5.3 miles from the original position of the submarine, 32 minutes after initial contact. The destroyer would have traveled a distance of 15.9 miles.

A reminder: When computing, make certain you express θ in radians, not degrees. Recall that 1.0 radian $= 360/2\pi = 57.3°$.

Examining Hailstone Numbers

After investigating two quite ancient sequences of numbers—the prime numbers and the Fibonacci numbers—we now turn to a sequence that is relatively new: the so-called hailstone numbers.

These numbers are generated by an extremely simple mathematical process in what is called the $3N + 1$ problem; it is also known as the Collatz problem and the Syracuse problem. By whatever name, it seems to have started in the 1930s, went away for a while, and then came back with renewed vigor in the early 1970s. For quite a few years it attracted the interest and efforts of many mathematicians in numerous American universities. Indeed, the joke went around in those days that the $3N + 1$ problem was planted in mathematics departments by enemy agents in a diabolical attempt to divert mathematicians from serious important research. To date, the basic problem has not been solved, so people are still working on it though perhaps not as vigorously as they were before.

In any event, here is a description of the $3N + 1$ problem and how it generates hailstone numbers. Select a positive integer, preferably small at the outset, for instance, less than 10 or 20. If it is an odd number, multiply it by three and add one (that's the $3N + 1$ thing); if it is even, divide it by two. Keep repeating this process until you cannot go any further. Let's try a few numbers and see what happens.

a. Start with $N = 3$. Then, applying the rules (if odd $3N + 1$; if even $N/2$) we generate the sequence 3, 10, 5, 16, 8, 4, 2, 1, 4, 2, 1, 4,

b. Start with $N = 5$. We obtain 5, 16, 8, 4, 2, 1, 4,

c. Start with $N = 7$. We obtain 7, 22, 11, 34, 17, 52, 26, 13, 40, 20, 10, 5, 16, 8, 4, 2, 1, 4, 2, 1,

So far, we have tried three starting numbers: 3, 5, and 7. In each case, our computations terminated when we entered a 4, 2, 1, 4, 2, 1 cycle. For the $N = 7$ case, we reached a *maximum value* of 52 and went through 16 computational steps (i.e., a *path length* of 16) before we reached the 421421 endless loop. Does this always happen regardless of the magnitude of the starting number? Perhaps we should begin with a larger number.

d. Start with $N = 25$. We obtain the sequence 25, 76, 38, 19, 58, 29, 88, 44, 22, 11, 34, 17, 52, 26, 13, 40, 20, 10, 5, 16, 8, 4, 2, 1. That's

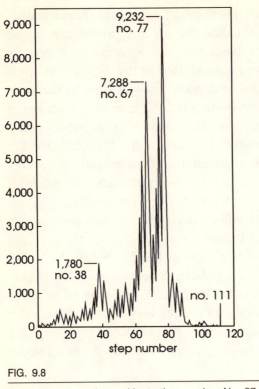

FIG. 9.8

The hailstone numbers with starting number $N = 27$.

more like it! This time we reached a maximum value of 88 and a path length of 23: an improvement but not good enough.

e. Start with $N = 27$. Well! What happened? As you will want to confirm, in this case the maximum value of the sequence was an amazing 9,232 and the path length was 111. A plot of this $N = 27$ sequence is shown in figure 9.8. We note that the plot displays several intermediate maximum values, notably 1,780 and 7,288, before it reaches the peak of 9.232 and then plunges to $N = 1$. Indeed, the entire plot of figure 9.8 looks like the great Wall Street crash of October 1929.

Incidentally, the reason these sequences are called "hailstone numbers" is because they behave something like the hail particles we associate with those enormous cumulonimbus clouds we call thunderheads. Small-diameter hail particles are caught in strong updrafts of air and are carried to great heights, growing in size as they do so. At high altitudes, these hail particles tumble out of the main current of the updraft and begin to fall, still growing in size. Near the base of the massive cloud they are again transported upward. This rising and falling pattern is repeated a few times. Finally, the hail particles become so large and heavy that they fall to the earth—something like our hailstone numbers going up and down several times before finally crashing to $N = 1$.

Up to here we have demonstrated that maximum values and path lengths of hailstone sequences depend somewhat on the magnitude of the starting number. More importantly, regardless of where the sequence starts, invariably it seems to get caught in the 421421 cycle and falls to $N = 1$.

We wonder if this is always the behavior. Does the sequence always terminate with 4, 2, 1, or could it, for some starting numbers and possibly after many ups and downs, nevertheless go off to infinity? Mathematicians have not yet been able to answer this question; the general solution remains an unsolved problem. Researchers at the University of Tokyo have tested every starting number up to 1.2 trillion; the sequence produced by every one of them sooner or later enters the 421 loop and ends up at $N = 1$. However, the conjecture that this always happens, regardless of the magnitude of the starting number, has not yet been theoretically established. For our general information, the paths lengths and successively larger maximum values of sequences corresponding to values of starting numbers N up to 100,000 are listed in table 9.3.

The hailstone number sequence: what a pleasant and easy problem to define and to carry out mathematically! Yet, so far, the $3N + 1$ conjecture has eluded rigorous proof by mathematicians. In the meantime, the problem nevertheless provides a very interesting exercise in arithmetic and point plotting for students at all levels.

Here are four references dealing with the $3N + 1$ problem and hailstone numbers. The first reference is the easiest: the paper by

TABLE 9.3

Starting numbers for sequences of hailstone numbers
and the corresponding maximum values and path lengths

Starting number	Path length	Maximum value
1	0	1
2	1	2
3	7	16
7	16	52
15	17	160
27	111	9,232
255	47	13,120
447	97	39,364
639	131	41,524
703	170	250,504
1,819	161	1,276,936
4,255	201	6,810,136
4,591	170	8,153,620
9,663	184	27,114,424
20,895	255	50,143,264
26,623	307	106,358,020
31,911	160	121,012,864
60,675	334	593,279,152
77,671	231	1,570,824,736

Source: Hayes (1984).

Bruce (1978) entitled "Crazy Roller Coasters." Then come Hayes (1984) and Lines (1990). The last reference, the most difficult, is by Lagarias (1985), with the title "The $3x + 1$ Problem and Its Generalizations." The author presents a short history of the problem and includes seventy references, most of which are not easy reading unless you're a real mathematician.

10

A Fast Way to Escape

Take a tennis ball and toss it straight upward. It goes to a certain height and falls back to your hands. Keep tossing or throwing the ball, each time with greater velocity. Observe that each time the ball goes higher and the total flight time is longer. An amazing observation! Anything else new? Most likely; we shall get to that shortly.

However, before going further we do magic: we remove all the air in the world. The resulting perfect vacuum greatly simplifies our mathematical analysis because now we can neglect the effect of air resistance. In this case, as we shall see, the height y_m to which our tennis ball ascends is $y_m = U_0^2/2g$, where U_0 is the velocity of the ball as it leaves your hand. The quantity g is the acceleration due to gravity; more precisely, it is the force exerted by gravity per unit mass of an object. In the metric system of units, g has the approximate value of 9.82 newtons of force per kilogram of mass or, equivalently, 9.82 meters per second squared; that is, $g = 9.82$ m/s^2. In the English or engineering system of units, $g = 32.2$ ft/s^2.

Now g is not really constant. At the earth's equator, $g = 9.780$ m/s^2 and at the poles, $g = 9.832$ m/s^2. The smaller value of g at the equator is due to the centrifugal force created by the rotation of the earth. As we shall see, g also depends very much on the distance from the center of the earth. However, for the moment, with a subscript to emphasize its constant value, we take $g_0 = 9.82$ m/s^2.

We begin our analysis with an introduction of Newton's famous laws of motion. These remarkable relationships were devised by the English mathematician, Issac Newton (1643–1727), one of the greatest mathematicians of all time. The three laws are stated as follows:

First law: An object at rest or moving at constant velocity will remain so unless an external force is applied

Second law: The acceleration of an object is proportional to the overall force applied on the object divided by the mass of the object

Third law: Every action (force) has an equal and opposite reaction (force)

Newton's second law is now employed in our tennis ball throwing experiment. We have

$$\sum F = ma, \tag{10.1}$$

in which $\sum F$ is the summation of all the forces acting on the ball, m is its mass, and a is its acceleration. By definition, we also have the relationships

$$a = \frac{dU}{dt} \text{ and } U = \frac{dy}{dt}. \tag{10.2}$$

These last two expressions indicate that the acceleration a is the rate of change of velocity U with time and that the velocity is the rate of change of vertical distance y with time.

If we neglect the effect of air resistance, the only force acting on the ball is its own weight. That is, $\sum F = -mg_0$. The minus sign is employed because the gravitational force acts in the downward (negative y) direction. So, utilizing equation (10.1) and the first relationship of (10.2), we obtain

$$\sum F = -mg_0 = m\frac{dU}{dt}, \tag{10.3}$$

which becomes

$$\frac{dU}{dt} = -g_0; \quad \int_{U_0}^{U} dU = -g_0 \int_{0}^{t} dt. \tag{10.4}$$

The second expression given in (10.4) is now in proper form for utilizing integral calculus. The quantities in the lower limits of the integrals describe the so-called "initial condition": $U = U_0$ when $t = 0$. Carrying out the integration yields

$$U = U_0 - g_0 t. \tag{10.5}$$

Utilizing the second relationship of (10.2) gives

$$\frac{dy}{dt} = U_0 - g_0 t; \quad \int_{0}^{y} dy = \int_{0}^{t} (U_0 - g_0 t) dt. \tag{10.6}$$

The lower limits of the integrals indicate that $y = 0$ when $t = 0$. Integrating equation (10.6) gives

$$y = U_0 t - \frac{1}{2} g_0 t^2. \tag{10.7}$$

Using equations (10.5) and (10.7) to eliminate the time variable t provides the relationship

$$U^2 = U_0^2 - 2g_0 y. \tag{10.8}$$

At the instant the tennis ball is at its highest point, $y = y_m$ and the velocity $U = 0$. So from (10.8), $y_m = U_0^2/2g_0$ or, alternatively, $U_0 = \sqrt{2g_0 y_m}$. For example, if the observed height $y_m = 10$ m, then the initial velocity $U_0 = 14.0$ m/s.

We note that if we neglect the effect of air resistance and assume that gravity is constant, it is easy to calculate the maximum height of our tennis ball. Clearly, the faster we throw the ball the higher it goes. However, we recall that the gravitational force g is not really constant. This leads us to another of Isaac

Newton's remarkable contributions: his famous law of gravitation,

$$F = G\frac{mM}{r^2},$$
(10.9)

in which F is the force of attraction between two bodies of masses m and M, whose centers are separated by a distance r. The quantity G is the gravitational constant. In the metric system of units, $G = 6.673 \times 10^{-11}$ newton m^2/kg^2.

We shall consider m to be the mass of the tennis ball and M the mass of the earth. If R is the radius of the earth and the force F is the weight of our ball at the earth's surface, then $F = mg_0$. Consequently, from equation (10.9) we obtain

$$F = mg_0\frac{R^2}{r^2},$$
(10.10)

where r is the distance of the ball from the earth's center. This equation says that the weight of the ball is inversely proportional to the square of its distance from the center of the earth. So it follows from equation (10.10) that $g = g_0 R^2/r^2$ describes the way in which g changes when $r > R$ (i.e., the ball outside the earth). Incidentally, $r = R + y$, where y, of course, is the vertical distance above the earth's surface. By the way, shortly we shall present an expression for g when $r < R$ (i.e., the ball is inside the earth). Analyses dealing with these matters are presented by Boas (1983).

Returning to Newton's second law, equation (10.1), we have

$$\sum F = -mg = m\frac{dU}{dt}.$$
(10.11)

Now, since $U = dr/dt$, we can write $dU/dt = (dU/dr)(dr/dt) = U(dU/dr)$. Also, since $g = g_0 R^2/r^2$, (10.11) becomes

$$\frac{g_0 R^2}{r^2} = -U\frac{dU}{dr}; \quad \int_{U_0}^{U} U\, dU = -g_0 R^2 \int_{R}^{r} \frac{dr}{r^2}.$$
(10.12)

The lower limits of the integral state that when $r = R$ (i.e., on the earth's surface), the "launch velocity" is U_0. Integrating this expression gives

$$U^2 = U_0^2 - 2g_0 R^2 \left(\frac{1}{R} - \frac{1}{r} \right). \tag{10.13}$$

Recall that when the gravitational force is constant, $g = g_0$, the velocity of the tennis ball becomes zero at a certain height $y_m = U_0^2/2g_0$. The ball then falls back to earth. However, when the gravitational force changes according to the relationship $g = g_0 R^2/r^2$, quite a different answer is obtained. We ask the question: can we toss the tennis ball or, more realistically, launch a projectile or space vehicle with sufficiently large velocity U_0 that it simply does not return? In other words, can the launch velocity be made so great that the ball or projectile actually escapes from the earth's gravity?

To answer the question, we certainly expect that when the distance (i.e., height) r of an object (a ball or projectile) gets larger and larger, its velocity U must approach zero. Accordingly, we set $r = \infty$ and $U = 0$ in equation (10.13). This yields the very simple expression

$$U_0 = \sqrt{2g_0 R}, \tag{10.14}$$

in which U_0 is the so-called "escape velocity." With $g_0 = 9.82 \text{ m/s}^2$ and $R = 6.37 \times 10^6$ m as the radius of the earth, we determine from this equation that $U_0 = 11,180$ m/s. Thus, if an object is propelled from the earth's surface at a velocity of 11.18 km/s or more, it will escape from the earth's gravitational field.

Incidentally, the escape velocity from the moon ($g_0 = 1.62 \text{ m/s}^2$ and $R = 1.74 \times 10^6$ m) is $U_0 = 2.37$ km/s. For the sun (with $g_0 = 274 \text{ m/s}^2$ and $R = 6.96 \times 10^8$ m), the escape velocity has the value $U_0 = 618$ km/s.

For the region $r > R$ (i.e., outside the earth) the gravitational force on an object is given by (10.10). By the same token, when

$r < R$ (i.e., inside the earth) the force is given by

$$F = mg_0 \frac{r}{R}.$$ (10.15)

This result is based on the assumption that the earth is homogeneous. It is well known that this is not precisely the case; however, for our analysis it is a reasonable assumption.

Analysis of Three Exciting Situations

Situation 1

In a scenario we leave to Jules Verne or other science fiction writers to explain, we are at the center of the earth ($r = 0$) and wish to launch our escape vehicle up an air-evacuated shaft, extending to the surface of the earth, with an initial velocity U_* sufficiently large just to reach the surface ($r = R$). What must be the magnitude of U_*?

From equation (10.15), we deduce that $g = g_0 r / R$. Substituting this into (10.11) and proceeding as before, we obtain

$$\frac{g_0 r}{R} = -U \frac{dU}{dr}; \quad \int_{U_*}^{U} U\, dU = -\frac{g_0}{R} \int_0^r r\, dr,$$ (10.16)

which gives

$$U^2 = U_*^2 - \frac{g_0}{R} r^2.$$ (10.17)

As required, $U = 0$ when $r = R$. Therefore, $U_* = \sqrt{g_0 R}$. Thus, a launch velocity $U_* = 7.91$ km/s is needed to get our vehicle from the center of the earth to the surface of the earth. A word of caution: at the instant of arrival at the surface, make sure that the vehicle is securely attached to a landing hook; otherwise, it will fall back down the shaft.

How long will it take to go from $r = 0$ to $r = R$? From equation (10.17), with $U_*^2 = g_0 R$ and using $U = dr/dt$, we have

$$U = \sqrt{\frac{g_0}{R}} \sqrt{R^2 - r^2} = \frac{dr}{dt}. \tag{10.18}$$

This equation can be expressed in the form

$$\int_0^{t_0} dt = \sqrt{\frac{R}{g_0}} \int_0^R \frac{dr}{\sqrt{R^2 - r^2}}, \tag{10.19}$$

where t_0 is the transit time. Utilizing a table of integrals we determine that $t_0 = (\pi/2)\sqrt{R/g_0}$. With $R = 6.37 \times 10^6$ m and $g_0 = 9.82$ m/s^2, the answer is $t_0 = 21$ minutes 5 seconds.

Situation 2

Alternatively, Jules Verne plans our escape not only from the evils raging at the center of the earth but also from those in great abundance at its surface. We shall be blasted off from $r = 0$ to a distant galaxy, infinitely far away. What initial velocity U_{*_0} is needed to accomplish this?

It should be clear that we can start with equation (10.17) with U_{*_0} replacing U_*. Also, we note that it is necessary to pass through the earth surface station at $r = R$ with the escape velocity, given by (10.14), of $U_0 = \sqrt{2g_0 R}$.

With this information, it is determined that $U_{*_0} = \sqrt{3g_0 R}$. This means that with a launch velocity $U_{*_0} = 13.70$ km/s, we can blast off from the center of the earth and go all the way to infinity. After the blast off, we pass through the earth surface station in 8 minutes 16 seconds, moving at a velocity of 11.18 km/s.

Situation 3

At the moment of our above-described departure from the center-of-the-earth launching station to go to infinity, a dreadful mistake is made. Instead of punching the $U_{*_0} = \sqrt{3g_0 R}$ blast-off velocity button, someone hits the $U_0 = \sqrt{2g_0 R}$ button. What happens?

PROBLEM   We leave the analysis of this situation to you as a homework problem. Here are the main results.

Our initial velocity, of course is 11.18 km/s.

We pass through the surface-of-the-earth station in 10 minutes 30 seconds at a velocity of 7.91 km/s. To our great dismay, this is much too slow to get us to infinity.

We reach a maximum height of $r = 2R$ (i.e., a distance of $R = 6,370$ km above the surface of the earth) 34 minutes 30 seconds after passing through the earth surface station or exactly 45 minutes after blast off.

After reaching maximum height, we begin our 34 minute 30 second fall back to the earth's surface. Theoretically, we will go back down the shaft and, after another 10 minutes 30 seconds, arrive at our $r = 0$ launching station with a velocity of 11.18 km/s.

Fortunately, just before we reenter the shaft, a friend turns the air back on. We bail out of our capsule, parachute to safety, and then quickly subdue the evils raging on earth.

11

How to Get Anywhere in About Forty-Two Minutes

Some rainy afternoon, when you have little to do, you might want to spend some time looking up the following rather interesting information. If you dug a hole straight downward from where you live, through the center of the earth, where would you come out?

If you live quite close to the intersection of Montana, Alberta, and Saskatchewan you are quite fortunate. You will come out on dry land, even though it is bleak and dreary Kerguelen Island in the south Indian Ocean.

If you live anywhere else in the continental United States, Alaska, or Canada, you'd better be ready to build a dike at the other end to keep the Indian Ocean out.

A dike is also needed if you live in Great Britain. The exit of your tunnel will be in the Pacific Ocean, south of New Zealand.

However, you are indeed lucky if you live in Honolulu. As you punch out the hole at the other end you will find yourself in west central Botswana in southern Africa.

It is good to know these things. They can be very useful for (a) initiating conversations, (b) changing the subject, and (c) eliminating awkward periods of silence at dinner parties, or even job interviews. They might also prove to be beneficial if you find yourself on *Jeopardy!* sometime.

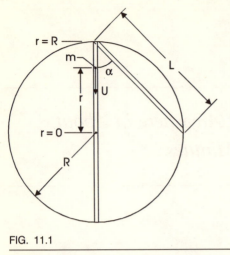

FIG. 11.1

Definition sketch for the motion of a vehicle in a shaft through the earth.

Well, after a few discussions and some serious thought, you and I decide there may be real commercial possibilities here. All we need is a well-constructed shaft of length equal to the world's diameter and a vehicle to carry passengers wishing to go from Hawaii to Botswana and vice versa. So we form a company, raise some cash, and hire some engineers and scientists to devise the necessary technology.

The first thing we need to do is carry out the basic analysis of the problem of our vehicle moving through the shaft. We decide to neglect air resistance, wall friction, centrifugal force due to the earth's rotation, and things like that. It is also assumed that the earth is a homogenous mass and that all thermal problems can be ignored. Actual conditions and circumstances involve details the engineers can work out.

A definition sketch of our problem is shown in figure 11.1. We start with the case in which the shaft passes through the center of the earth (i.e., $\alpha = 0°$). Recall that Newton's second law of motion is

$$\sum F = ma, \tag{11.1}$$

in which $\sum F$ is the summation of all the forces acting on the vehicle, m is its mass, and a its acceleration. Neglecting air resistance, wall friction, and centrifugal force, the only force involved is the weight of the vehicle, that is, $\Sigma F = mg$.

As we saw in chapter 10, the gravitational force g inside the earth is given by the expression $g = g_0 r / R$, where $g_0 = 9.82 \ \text{m/s}^2$ is the gravitational force at the earth's surface, r is the distance from the center of the earth, and $R = 6.37 \times 10^6$ m is the radius of the earth. In addition, we have the relationships

$$U = \frac{dr}{dt} \text{ and } a = \frac{dU}{dt}, \tag{11.2}$$

where t is time, U is the velocity of the vehicle, and a is its acceleration. The weight of the vehicle, acting in the negative r direction, is $-mg_0 r / R$.

Utilizing these relationships, we obtain the expression

$$\frac{d^2 r}{dt^2} + \frac{g_0}{R} r = 0. \tag{11.3}$$

This is a simple example of a so-called differential equation. We seek its solution, $r = f(t)$. An excellent reference for problems of this kind is the book by Boyce and DiPrima (1996). They provide the following solution to this equation:

$$r = c_1 \cos \omega t + c_2 \sin \omega t, \tag{11.4}$$

where c_1 and c_2 are constants and $\omega = \sqrt{g_0 / R}$ is the so-called *frequency* of oscillation. The values of c_1 and c_2 are determined from the "initial conditions": (a) $r = R$ when $t = 0$ and (b) $U = 0$ when $t = 0$. It is easily established that $c_1 = R$ and $c_2 = 0$. Accordingly, the answer is

$$r = R \cos \frac{2\pi}{T} t, \tag{11.5}$$

in which $T = 2\pi / \omega = 2\pi \sqrt{R / g_0}$ is the *period* of oscillation. This equation describes *simple harmonic motion*.

Substituting the numerical values of R and g_0 into this expression, it is determined that the period of oscillation is $T = 5{,}060$ seconds $= 84$ minutes 20 seconds. This is the time needed for a complete oscillation, that is, the round trip of our vehicle. The duration of the one-way journey is $T/2 = 42$ minutes 10 seconds.

The following expressions for the velocity U and acceleration a are obtained by using equations (11.2) and (11.5):

$$U = \sqrt{g_0 R} \, \sin \frac{2\pi}{T} t; \quad a = g_0 \cos \frac{2\pi}{T} t. \tag{11.6}$$

It is observed that the maximum acceleration (or deceleration) is g_0 (or $-g_0$) and occurs at $t = 0$ (or $T/2$). The maximum velocity is experienced when the vehicle passes through $r = 0$ at $t = T/4$; its value is $U_m = \sqrt{g_0 R} = 7{,}909$ m/s. The average velocity during the journey is $\overline{U} = (2/\pi)\sqrt{g_0 R} = 5{,}035$ m/s.

This is really fast! In fact, assuming that the speed of sound in air is $C = 340$ m/s, we will be moving at about Mach 23 as we pass through the center of the earth. But do not forget that there is no air in the shaft.

In any event, if a tour group from Botswana wants to spend a surfboarding weekend in Hawaii, they can get there in about 42 minutes. Likewise, if people want to escape the high cost of living in Honolulu, they can be in Botswana after a 42-minute trip—not even time for a nice nap.

Startling Piece of Information 1

It turns out that the period of oscillation, $T = 2\pi\sqrt{R/g_0}$, is the answer to our problem whether or not the shaft passes through the center of the earth. With reference to figure 11.1, suppose that the shaft is inclined at some angle α connecting any two points on the earth's surface. The period is still the same and, regardless of the angle α, the one-way journey is always 42 minutes 10 seconds.

On this point we shall not go through the details of the analysis; however, you may want to do the computations. Suffice it to say that the length of the shaft is $L = 2R \cos \alpha$ and so the journey is shorter. However, as we could show, the average velocity, $\overline{U} = (2/\pi)\sqrt{g_0 R} \cos \alpha$, is less because the average value of the gravitational force is less. Accordingly, with $T = L/\overline{U}$, the cosine terms cancel and the period is the same as before.

Startling Piece of Information 2

For this we need Newton's law of gravitation, which is given by the equation

$$F = G \frac{mM}{r^2}, \qquad (11.7)$$

where m is the mass of our vehicle, M is the mass of the earth, r is the distance between the centers of gravity of the two masses, and $G = 6.673 \times 10^{-11}$ newton m^2/kg^2 is the gravitational constant. Now at the surface of the earth ($r = R$), the force exerted by gravity on the vehicle is $F = mg_0$. Substituting this relationship into (11.7) gives $g_0 R^2 = GM$. Multiplying both sides of this expression by $4\pi R/3$ and noting that the density of the earth, by definition, is $\rho = M/V$, where $V = 4\pi R^3/3$ is the volume, we obtain the relationship,

$$\frac{R}{g_0} = \frac{3}{4\pi} \frac{1}{G\rho}. \qquad (11.8)$$

Consequently, with $T = 2\pi\sqrt{R/g_0}$, we establish the identity,

$$T = \sqrt{\frac{3\pi}{G\rho}}, \qquad (11.9)$$

or, in MKS units, $T = (3.758 \times 10^5)/\sqrt{\rho}$ where ρ is the density of the earth in kg/m^3.

These two startling pieces of information tell us that the period of oscillation, T, (a) is independent of the angle of the shaft, α, and (b) depends only on the density, ρ.

Table 11.1 lists some data and computed results involving the eleven well-known bodies of our solar system. It is interesting to note that the earth has the largest density of any body in the

TABLE 11.1

Motion of an object in a shaft
Values of various parameters for the bodies of the solar system.

Body	Density ρ, kg / m^3	Radius R, m	Surface gravity g_0, m / s^2	Period of oscillation T		Maximum velocity U_m, km / s
Sun	1,410	6.96×10^8	274.27	166m	4s	436.91
Moon	3,330	1.74×10^6	1.62	108m	3s	1.68
Mercury	5,420	2.44×10^6	3.78	85m	5s	3.04
Venus	5,090	6.05×10^6	8.60	87m	50s	7.21
Earth	5,520	6.37×10^6	9.82	84m	20s	7.91
Mars	3,920	3.40×10^6	3.72	100m	7s	3.56
Jupiter	1,140	71.90×10^6	22.88	185m	38s	40.56
Saturn	700	57.50×10^6	11.25	236m	45s	25.43
Uranus	1,060	26.15×10^6	7.77	192m	7s	14.25
Neptune	1,590	24.75×10^6	11.00	157m	5s	16.50
Pluto	4,400	3.50×10^6	4.30	94m	29s	3.88

Source: Audouze and Israël (1985).

solar system, although the densities of Mercury and Venus are only slightly less. By a substantial margin, Saturn's density is the smallest. These densities are reflected in the computed periods of oscillation, T, shown in the table. Although there may be numerous disadvantages to living on planet earth, at least we have the shortest transglobal commuting time anywhere in the solar system.

Values of the maximum velocity, $U_m = \sqrt{g_0 R}$, are also listed in table 11.1. We recall from chapter 10 that the so-called "escape velocity" is $U_0 = \sqrt{2 g_0 R}$. The relationship between the two velocities is interesting.

Example 1: Falling Through a Bottomless Well

This problem of a shaft through the center of the earth is discussed by Perelman (1982) in his book *Physics Can Be Fun*. Essentially, he describes a case in which a "bottomless well" extends from a high plateau in eastern Ecuador (elevation

2,000 m), passes through the center of the earth, and comes out at sea-level Singapore.

To elaborate on Perelman's scenario, a foreign agent jumps into the Ecuadorian end of the shaft and, about 42 minutes later, comes sailing out of the Singapore exit at a velocity of around 715 km/hr. He peaks out at a height of 2,000 m above Singapore, hastily takes video movies of the harbor defense works, and, after a fall of approximately 20 seconds, disappears into the shaft. Following another 42-minutes journey he is back in Ecuador. He climbs out of the shaft and—his mission accomplished—vanishes.

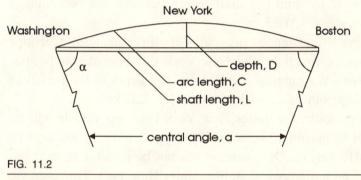

FIG. 11.2

Dimension diagram of Northeast Corridor gravity-driven maglev transportation system.

Example 2: A High-Speed Transportation System

Instead of trying to construct a shaft connecting Ecuador and Singapore or Honolulu and Botswana, we decide to start with something less complicated: a high-speed transportation system featuring a tunnel connecting Washington, D.C. and Boston.

First, what is the length of such a Northeast Corridor tunnel? With reference to figure 11.2, we can calculate the length C of the arc of a great circle from the equation

$$\cos a = \sin \phi_1 \sin \phi_2 + \cos \phi_1 \cos \phi_2 \cos(\lambda_1 - \lambda_2),$$

(11.10)

in which ϕ_1, λ_1 are the latitude and longitude of Washington and ϕ_2, λ_2 are those of Boston. The angle a is the arc angle defined in the figure. Clearly, $C = (a/360)(2\pi R)$ where R is the radius of the earth. We obtained (11.10) from the section on spherical trigonometry in the comprehensive mathematics reference book by Gellert et al. (1977).

The following information concerning the coordinates of Washington and Boston is obtained from *The World Almanac* (1994): Washington, longitude $\lambda_1 = 77°01'$W, latitude $\phi_1 = 38°54'$N; Boston, longitude $\lambda_2 = 71°03'$W, latitude $\phi_2 = 42°21'$N. Substitution of these coordinates into (11.10) gives $a = 5.68°$ and consequently $C = 632$ km. It is easily determined that the shaft angle $\alpha = 87.16°$, and the shaft length between the two cities is $L = 2R \cos \alpha = 631.23$ km.

By sheer coincidence, the path of our Northeast Corridor tunnel passes directly under New York City approximately midway between Washington and Boston. The depth of the tunnel at this midway point is $D = R(1 - \sin \alpha) = 7.82$ km.

As we pass directly under New York City, our vehicle will be moving at its maximum velocity, $U_m = \sqrt{g_0 R} \cos \alpha = 392$ m/s or about 1,410 km/hr. Of course, it will not be possible to stop the vehicle at an underground station under New York. However, we could drop off some mail—assuming we have a sturdy mail catcher. The average velocity for the entire trip, $\overline{U} = 2v_m/\pi = 250$ m/s = 898 km/hr. The duration of the trip, of course, is $T = L/\overline{U} = 42$ minutes 10 seconds.

The journey on our high-speed vehicle will be pleasant, though perhaps a bit dull. After activating the magnetic levitation cradle and passing through the air chamber seals, our vehicle begins its trip by going down a mild slope of $a/2 = 2.84°$. The acceleration, $a = (g_0 \cos \alpha) \cos \omega t$, is initially 0.487 m/s^2 (which corresponds to about 11 miles per hour in 10 seconds); this is the maximum acceleration.

Large vacuum pumps keep air out of the tunnel (to eliminate air resistance). The magnetic levitation system suspends the vehicle (to eliminate rail resistance); it is also designed to counteract Coriolis forces caused by the earth's rotation. Tunnel walls are insulated to eliminate heat and moisture intrusion.

It will be necessary to provide power for the vacuum pumps and for the magnets. For the latter, power will be minimized with the use of cryogenic superconducting materials.

However, since our maglev vehicle is "falling downhill" due to the force of gravity and then "coasting uphill," there will be no power costs for propulsion. Further, there will be no costs associated with climate and weather and other environmental factors. Finally, since the duration of the journey is only about 42 minutes, there will be no costs involved in providing movies, headsets, or expensive meals for the passengers. Our profit should be large.

12

How Fast Should You Run in the Rain?

There is a steady downpour of rain, you have no raincoat and no umbrella, and you must get from here to there without delay. And instinctively, you want to get the least wet as you make the excursion of specified distance.

If you walk slowly in the rain, only the top of your head and shoulders get wet; your front stays relatively dry as do your shoes and socks. However, with a slow walk you are out in the rain that much longer. If you jog at moderate speed, your front gets wet and your shoes get soggy from splashing water on the pavement; however, in this case your trip time is reduced. And if you run at full speed, the top of your head and shoulders avoid most of the rain, your front is drenched and your shoes and socks are entirely waterlogged. Even so, in this instance the time you are out in the rain is least.

So the question is asked: with what speed should you move to get least wet? This problem has been examined by Rizika (1950) and by Edwards and Hamson (1990). Both of these earlier analyses included the effect of nonvertical rainfall due to wind but disregarded pavement splashing. We shall do the reverse: neglect wind action but include splashing. This illustration demonstrates the usefulness of differential calculus in solving problems involving maxima and minima.

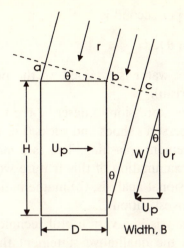

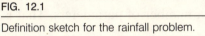

FIG. 12.1

Definition sketch for the rainfall problem.

Analysis of the Problem

The basic dimensions and units of our problem are the following: mass of water (kilograms), distance (meters), and time (seconds). The main variables are a person's velocity U_p (m/s), raindrop velocity U_r (m/s), and rainfall rate r (m/s or mm/hr). To characterize the person's dimensions we select an esthetically unattractive but mathematically simple rectangular prism with height H, width B, and thickness D. The person moves in a direction parallel to the D dimension, that is, perpendicular to the BH plane.

A definition sketch is presented in figure 12.1. We note that the combination of velocities U_p and U_r establishes a resultant velocity $W = \sqrt{U_p^2 + U_r^2}$, inclined at an angle θ to the BH plane.

The volume of rain crossing plain ab per second is

$$q_1 = rB(D \cos \theta). \tag{12.1}$$

This is the amount of water collected on the person's head and shoulders (i.e., the top of the prism). Similarly, the volume of rain crossing plane *bc* per second is

$$q_2 = rB(H \sin \theta), \tag{12.2}$$

and this amount of water is retained on the person's front (i.e., the front of the prism).

Next we need an expression to describe the volume of water q_3 collecting on the person's shoes and socks (i.e., the bottom of the prism) due to pavement splashing. This is a more difficult problem. A complete examination of this feature would involve three aspects: (a) dimensional analysis, (b) mathematical modeling, and (c) experimental investigation.

Dimensional analysis is a very useful technique that starts, in our problem, with the qualitative statement that the volume of splashing water may depend on a number of parameters. That is, we suppose that

$$q_3 = f(d, B, D, U_p, \rho, \mu, \sigma, g). \tag{12.3}$$

This expression indicates that the magnitude of q_3 is probably affected by the depth of water on the pavement (d), by the bottom dimensions (B and D), by the person's velocity (U_p), and by the rainwater's mass density (ρ), dynamic viscosity (μ), surface tension (σ), and gravitational force (g).

A dimensional analysis of the variables listed in equation (12.3) yields the relationship

$$q_3/(BDU_p) = f(d/D, B/D, \text{Re}, \text{We}, \text{Fr}), \tag{12.4}$$

where Re, We, and Fr are, respectively, the so-called Reynolds, Weber, and Froude numbers. These quantities are very important in the field of fluid mechanics. Barenblatt (1987) is an easy-to-understand book that deals with all of these topics.

Augmenting the techniques of dimensional analysis are straightforward mathematical analysis and experimental studies. With regard to our rainfall problem, reference is made to the experiments of Mutchler and Larson (1971) concerning the mechanics of raindrop splash erosion of soil. Utilizing their results, as well as those of dimensional and mathematical analyses, we establish that the volume of water splashing onto the bottom plane BD per second is

$$q_3 = \frac{cd}{\nu} BDU_p^2,$$ \hfill (12.5)

where c is a dimensionless constant, d is the water depth on the pavement, and $\nu = \mu/\rho$ is the kinematic viscosity of water.

Utilizing the relationships $\cos\theta = U_r/W$ and $\sin\theta = U_p/W$, we combine equations (12.1), (12.2), and (12.5) to obtain

$$q = rBD\frac{U_r}{W} + rBH\frac{U_p}{W} + \frac{cd}{\nu}BDU_p^2,$$ \hfill (12.6)

which expresses the total volume of water collected on the person per second. The total number of seconds during the journey of length L is $T = L/U_p$. Using $W = \sqrt{U_p^2 + U_r^2}$, the total volume of water collected on the trip is

$$V = \frac{rBL}{\sqrt{U_p^2 + U_r^2}}\left(D\frac{U_r}{U_p} + H\right) + \frac{cd}{\nu}(BDL)U_p.$$ \hfill (12.7)

This equation can be put into the following dimensionless form:

$$Z = \frac{1}{\sqrt{1 + S^2}}\left(\frac{n}{S} + 1\right) + fnS,$$ \hfill (12.8)

in which $Z = (VU_r)/(BHLr)$ is the wetness factor, $S = U_p/U_r$ is the speed ratio, $n = D/H$ is the aspect ratio, and finally $f = cdU_p^2/\nu r$ is the splash number.

A word about this last parameter, the splash number. This is simply an index or scale to describe how wet the pavement or ground surface is. In this sense, it is like the Dow–Jones index for the stock market, the Richter scale for earthquakes, or the Beaufort number for a sea-wind condition. A short list of splash numbers is presented in table 12.1.

The main result of our analysis is equation (12.8). This expression indicates that the amount of wetness, Z, depends on the person's speed as a fraction of raindrop velocity. Several plots of (12.8) are shown in figure 12.2 for various values of splash number f, with a constant value of aspect ratio, $n = 0.15$.

TABLE 12.1

Splash number *f* for various surfaces

Surface description	Value of *f*
Dry pavement or dry grass	0
Damp grass	5
Wet pavement	10
Puddly pavement	15
Gutter or very shallow stream	20

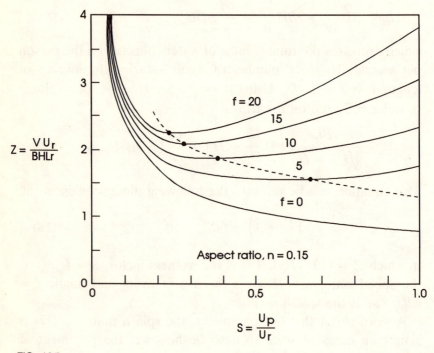

FIG. 12.2

Wetness factor *Z* as a function of the speed ratio *S* for various values of the splash number *f*.

In the figure we note several things. First, if $f = 0$, that is, the pavement or grass is dry, there is no optimum point on the curve. In this case, a person should run as fast as possible to minimize the amount of rain collected. However, for values of $f > 0$ there are optimum speed points that do provide minimum wetness. Furthermore, as f increases, the optimum speed decreases. In this case, a person should adjust his or her speed in order to minimize wetness.

Now one could determine the "best" speed ratio directly from figure 12.2. However, if we utilize the methods of differential calculus we can be more precise. This involves a step known as "taking the first derivative" of equation (12.8). Basically, this procedure allows us to compute the "slope" at any point on one of the curves of figure 2.12. We note that when the slope is zero (i.e., a line parallel to the S-axis), we have identified the value of S corresponding to the minimum value of Z. Of course, this is what we are after.

If we take the first derivative of equation (12.8) and set it equal to zero, we obtain

$$S_m^3 + 2nS_m^2 + n = fnS_m^2\left(1 + S_m^2\right)^{3/2}, \tag{12.9}$$

where S_m is the value of the speed ratio corresponding to the minimum value of the wetness factor, Z_m. Now in a great many problems involving determination of minimum values it is possible to solve an equation analogous to (12.9) to obtain the desired quantity. In our problem, however, we cannot obtain an explicit solution for S_m. Nevertheless, for specified values of f and n, it is not difficult to compute S_m from equation (12.9) by trial-and-error calculation. Table 12.2 lists the minimum points for the curves displayed in figure 12.2. The dashed line in the figure is the locus of these points.

There is considerable information about the velocity of raindrops. A typical study is that of Dingle and Lee (1972), who show that the terminal velocity of a drop depends primarily on its diameter. A list of raindrop velocities at sea level, computed from their equation, is shown in table 12.3.

TABLE 12.2

Optimum velocity coordinates

f	S_m	Z_m
0	—	—
5	0.668	1.519
10	0.363	1.873
15	0.278	2.110
20	0.235	2.300

TABLE 12.3

Raindrop velocities at sea level

Diameter, mm	U_r, m/s
1.0	3.9
2.0	6.6
3.0	8.1
4.0	8.8
5.0	9.1

Of course, in a particular rainstorm there is a wide distribution of raindrop diameters and hence raindrop velocities. If the rainfall rate $r = 25$ mm/hr, the mean diameter is about 1.9 mm, and if $r = 100$ mm/hr, the mean drop size is approximately 2.5 mm. Accordingly, a typical raindrop velocity is around 6 to 7 m/s.

An Example

A young lady leaves her office at the end of the day to face her customary ($L = 500$ m) trip to the train station. Unfortunately, there is a pouring-down rain. Indeed, it has been raining all afternoon; the soggy route to the station can best be characterized as "puddly pavement" ($f = 15$).

The young lady's dimensions, brutally simplified, are $B = 55$ cm, $D = 24$ cm, and $H = 160$ cm; accordingly, $n = D/H = 0.15$. Before she steps out into the rain to begin her trek, she quickly ascertains that the rainfall rate is $r = 25$ mm/hr and the raindrop velocity is $U_r = 6.5$ m/s.

Then she makes the following rapid calculations. From table 12.2 she determines that $S_m = 0.278$ and $Z_m = 2.110$. Since $S_m = U_{p,m}/U_r$ and $U_r = 6.5$ m/s, she easily computes that her jogging speed to get least wet should be $U_{p,m} = 1.81$ m/s $= 5.93$ ft/s. At this speed she will be able to reach the train station in $T = 500/1.81 = 276$ seconds $= 4.6$ minutes.

As she jogs to the station, she remembers the first of the relationships following equation (12.8):

$$Z = \frac{V}{BH}\frac{U_r}{Lr} \text{ or } V = \frac{BHLr}{U_r}Z. \tag{12.10}$$

Making certain that all quantities are in consistent (MKS) units, she substitutes the numbers into (12.10):

$$V = \frac{(0.55)(1.60)(500)(6.94 \times 10^{-6})}{(6.5)}(2.110), \tag{12.11}$$

to determine that $V = 991 \times 10^{-6}$ m^3 $= 991$ cm^3 of water has collected or will collect on her person during her jog. The weight of this water is 991 grams or about 2.19 pounds. Finally, from the relative values of the terms of equation (12.8) she determines that 224 grams (24.6% of the total) of water collects on her head and shoulders, 453 grams (45.7%) on her jacket and slacks, and 294 grams (29.7%) on her shoes and socks.

This point concerning distribution of collected water is illustrated in figure 12.3 for the case $f = 15$ and $n = 0.15$. We note from the figure that the Z_1 and Z_3 curves descend and ascend, respectively, with increasing speed ratio S. However, the Z_2/Z (frontal wetness) curve possesses a maximum point. To determine this point we simply construct the ratio of the second term of equation (12.8) to the sum of the three terms. Then, as before, we take the first derivative of the resulting expression and equate it to zero. The answer is

$$S_*^2\left(S_*^2 + \frac{1}{2}\right) = \frac{1}{2f}(1 + S_*^2)^{1/2}, \tag{12.12}$$

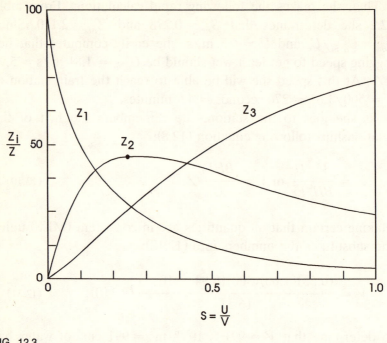

FIG. 12.3

Percentage distribution of collected rainfall. Top plane, Z_1, frontal plane, Z_2, bottom plane, Z_3. Splash number $f = 15$; aspect ratio $n = 0.15$.

in which S_* is the speed ratio corresponding to the maximum value of Z_2/Z. In our example, $S_* = 0.247$, $(Z_2/Z)_* = 0.459$, and $Z_* = 2.116$. We note that the critical point (S_*, Z_*) is very close to the more important critical point (S_m, Z_m).

Back to our clever young lady. Clearly, if she is wearing a new hat or has a new hairdo she should, according to the Z_1/Z curve of figure 12.3, run as fast as possible. Alternatively, if she has a new pair of shoes then, from the Z_3/Z curve, she should walk quite slowly. Finally, if she has a new hat, a new pair of shoes, and a new jacket and slacks, she should take a taxi to the station.

13

Great Turtle Races:
Pursuit Curves

We have some small turtles that are very intelligent and well trained; they are what we call smart turtles. They can be ordered to stay put, to walk or trot along a straight line or some curved path, and to follow other specific instructions. They refuse to swim and they cannot fly; hence their movements are restricted to zero, one, or two dimensions, that is, to a point, a straight line, or a plane, respectively.

With that as a prologue, we examine what is termed the *curve of pursuit problem*. This is quite an ancient problem in mathematics. Evidently it was first studied by Leonardo da Vinci in the early sixteenth century, and it has attracted the attention of mathematicians ever since.

With reference to the definition sketch of figure 13.1, we have two curves in the *x-y* plane. Curve *A* is the path of the so-called *pursuer* and curve *B* that of the *pursued*. That is, at any particular moment the pursuing turtle (or frigate, cheetah, bad guy on horseback) is at point $P(x, y)$ and moves along path *A* at a velocity *u*. At the same moment, the pursued turtle (or pirate ship, gazelle, stagecoach) is at point $Q(m, n)$ and moves along path *B* at a velocity *v*.

The curve of pursuit problem follows: if the pursuer *P* always heads directly toward the pursued *Q*, and if the path *B* of the pursued is specified, then what is the path *A* of the pursuer?

CHAPTER 13

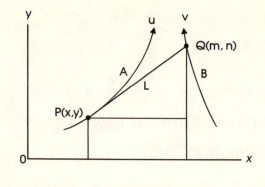

FIG. 13.1

Definition sketch for the curve of pursuit problem.

The key phrase is "... always heads directly toward... ." This means that the slope of the pursuit curve is

$$\frac{dy}{dx} = \frac{n - y}{m - x},$$

(13.1)

where dy/dx is the slope of curve A. The first derivative of this equation yields

$$\frac{dn}{dx} = (m - x)\frac{d^2y}{dx^2} + \frac{dm}{dx}\frac{dy}{dx}.$$

(13.2)

In addition, we have the relationship $u = kv$, where u and v are the velocities of the pursuer and pursued; k is a constant which may be either larger or smaller than one. Since $u = ds/dt$ and $v = d\sigma/dt$, where ds and $d\sigma$ are incremental distances along the two curves, we obtain

$$(dx^2 + dy^2) = k^2(dm^2 + dn^2),$$

(13.3)

which can be written in the form

$$1 + \left(\frac{dy}{dx}\right)^2 = k^2\left\{\left(\frac{dm}{dx}\right)^2 + \left(\frac{dn}{dx}\right)^2\right\}.$$

(13.4)

Finally, we assume that the path of the pursued, $m = g(n)$, is a known relationship. This information, along with equations (13.2) and (13.4), gives us what we want: the path of the pursuer, $y = f(x)$.

Sounds easy? Not always. Even with some substantial simplifications already (e.g., a plane surface and constant velocities) this can be a difficult problem. For example, suppose we have a duck paddling at constant speed v around the periphery of a large circular pond. A small alligator, at the center of the pond at time $t = 0$, sights the duck and swims with velocity u directly toward the tasty meal. Under what conditions will the alligator catch the duck, and what will be the alligator's path?

It turns out that this is not a simple problem. In fact, as Davis (1962) and Melzak (1976) show, an exact answer is impossible though, of course, numerical and graphical solutions are available. As we might expect, when $k > 1$ (the alligator swims faster than the duck) capture occurs. However, when $k < 1$ the duck is safe; the frustrated alligator simply swims around in a circle, a so-called "limit cycle," of radius kR, where R is the radius of the pond.

We go to an easier problem. As shown in figure 13.2, two turtles are situated on a horizontal plane. Initially turtle A, the pursuer, is located at the origin $(0, 0)$ and is facing east. Turtle B, the pursued, is at $(a, 0)$ and is facing north. At $t = 0$, the turtles commence the race. Pursued turtle B has instructions to proceed at constant velocity v along the straight line $m = a$. Accordingly, $dm/dx = 0$ and (13.2) becomes

$$\frac{dn}{dx} = (m - x)\frac{d^2y}{dx^2}, \tag{13.5}$$

and so from (13.4) we obtain

$$\sqrt{1 + \left(\frac{dy}{dx}\right)^2} = k(m - x)\frac{d^2y}{dx^2}. \tag{13.6}$$

This differential equation is solved to provide the answer we want, $y = f(x)$.

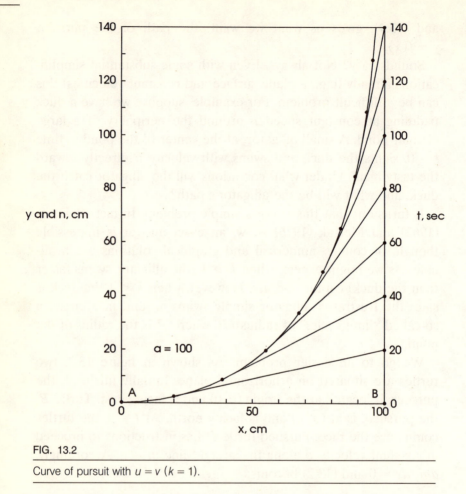

FIG. 13.2

Curve of pursuit with $u = v$ ($k = 1$).

For the first race the two turtles are ordered to move at the same velocity, that is, $u = v$, and so $k = 1$. In this instance, the solution to equation (13.6) is

$$y = \frac{a}{4}\left\{\left(1 - \frac{x}{a}\right)^2 - \log_e\left(1 - \frac{x}{a}\right)^2 - 1\right\}. \tag{13.7}$$

This equation describes the path of pursuing turtle A and is displayed in figure 13.2 with $a = 100$ cm and $u = v = 1.0$ cm/s. Not surprisingly, turtle A never catches up with turtle B. However, we note that at the start of the race there was a distance

$a = 100$ cm between them. Because turtle A took the indicated shortcut, he eventually closed the gap to 50 centimeters—but no more.

For all ensuing races, the turtles are instructed to move at speeds such that $k \neq 1$. In this case, we obtain the following solution to equation (13.6):

$$y = \frac{ka}{(k^2 - 1)} + \frac{ka}{2(k^2 - 1)} \left\{ (k - 1)\left(1 - \frac{x}{a}\right)^{1 + 1/k} \right.$$
$$\left. - (k + 1)\left(1 - \frac{x}{a}\right)^{1 - 1/k} \right\}. \tag{13.8}$$

If $k < 1$, turtle A will never capture turtle B. However, if $k > 1$, capture will always result; this will occur when $x = a$. So from (13.8) we immediately have

$$y_c = \frac{ka}{(k^2 - 1)}, \tag{13.9}$$

where y_c is the capture point. For example, if $k = 1.5$ then $y_c = 6a/5$; if $k = 2$, $y_c = 2a/3$; and if $k = 3$, $y_c = 3a/8$.

Reversing the question, with what speed must turtle A move to capture turtle B at a specified y_c? Solving equation (13.9) for k gives

$$k = \frac{a}{2y_c}\left(1 + \sqrt{1 + \frac{4y_c^2}{a^2}}\right). \tag{13.10}$$

Suppose we specify that $y_c = a$. Then from this equation we get $k = (1/2)(1 + \sqrt{5}) = 1.61803$.

What a nice surprise. Once again, we run across one of history's truly famous numbers. You will recall we studied it in an earlier chapter: the golden number or golden ratio, $\phi = (1/2)(1 + \sqrt{5}) = 1.61803$.

One thing can be said with certainty about ϕ: it shows up in the strangest places. Its appearance here as the speed ratio, $k = \phi$, in our turtle pursuit problem indeed demonstrates the amazing characteristics of ϕ to show up unexpectedly. Numerous

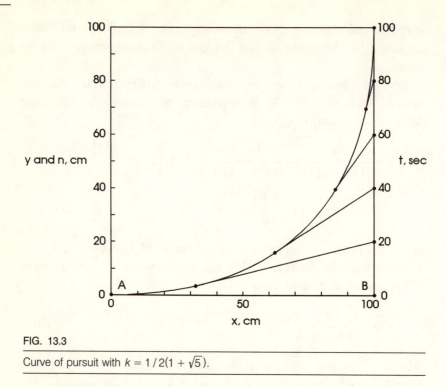

FIG. 13.3

Curve of pursuit with $k = 1/2(1 + \sqrt{5})$.

other remarkable appearances and features of this fascinating number are given by Huntley (1970) and by Kappraff (1990).

A plot of this particular pursuit curve is shown in figure 13.3. The capture point is $y_c = a = 100$ cm and turtle B's speed is $v = 1.0$ cm/s. Turtle A, with speed $u = kv = 1.618$ cm/s, covers a distance of $S = 161.8$ cm in a path that is almost circular. The precise path, of course, is determined by substituting $k = (1/2)(1 + \sqrt{5})$ into equation (13.8).

In the preceding pursuit curve analysis we stipulated *constant velocities* of the pursuer and the pursued. A closely related problem is one that stipulates *constant distance* between the two.

As before, equation (13.1) applies. The pursuer always heads directly toward the pursued. However, this time we impose the condition of constant distance:

$$L = \sqrt{(m - x)^2 + (n - y)^2},$$

(13.11)

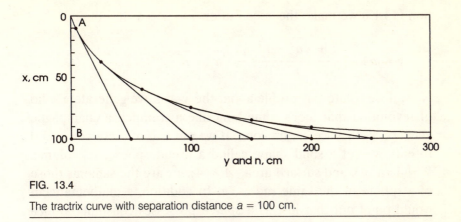

FIG. 13.4

The tractrix curve with separation distance $a = 100$ cm.

where, as indicated in figure 13.1, L is the required distance between $P(x, y)$ and $Q(m, n)$. Again, the path of the pursued, $m = g(n)$, is specified. This information, along with equations (13.1) and (13.11), determines the path of the pursuer, $y = f(x)$.

We consider the same relatively easy problem as before: $m = a$. In this case,

$$a = \sqrt{(a - x)^2 + (n - y)^2}.$$

(13.12)

Solving for $(n - y)$, substituting into (13.1), and integrating gives the solution

$$y = a \log_e \frac{a + \sqrt{a^2 - (a - x)^2}}{(a - x)} - \sqrt{a^2 - (a - x)^2}.$$

(13.13)

This equation is called the tractrix; its shape is shown in figure 13.4. Recall that we ordered turtle A, starting at the origin, always to maintain a constant distance a from turtle B as the latter moves along. However, the same problem would be presented had we simply instructed turtle B to attach a string to turtle A and drag him along.

The tractrix has numerous interesting properties and these are discussed by Lockwood (1961) and by Steinhaus (1969). In equation (13.13) we make the substitution $r = a - x$, to obtain

$$y = a \log_e \frac{a + \sqrt{a^2 - r^2}}{r} - \sqrt{a^2 - r^2} \,. \tag{13.14}$$

Now, if we rotate this profile about the y-axis, we generate a solid of revolution that looks something like a trumpet, a champagne glass, or a fancy ice cream cone. If we put two of these cones end to end, we get a solid shape called a pseudosphere. Its volume, $V = 4\pi a^3/3$, and surface area, $A = 4\pi a^2$, are the same as those of a sphere of the same radius, a. In addition, a mathematician would point out that just as a *sphere* possesses constant *positive* curvature (and hence serves as a model for non-Euclidean elliptic geometry), the *pseudosphere* has constant *negative* curvature (and so serves as the basis for non-Euclidean hyperbolic geometry).

Our two turtles are probably much more interested in the race. We shall bet on turtle *B*.

14

More Great Turtle Races: Logarithmic Spirals

More turtle racing: This time we shall use several or many turtles and instruct all of them to move at the same speed. So it is not really racing; as we shall see, it is more a matter of symmetrical pursuit.

We start with a configuration involving two turtles. Turtle A is located at the west end of a straight line of length a, facing east. Turtle B is at the east end facing west. When the starting gun is fired, each turtle heads directly toward the other. Obviously, the two turtles will collide at the midpoint of the line and each will have covered a distance $a/2$.

Since that was not too difficult we move on to the three-turtle problem. This is not as easy. As shown in figure 14.1, initially the turtles are at the corners of an equilateral triangle with side length a. Turtle A is facing turtle B who is facing turtle C who is facing turtle A. Again, as the race commences, each turtle always heads directly toward his target.

The salient features of our analysis are that the three turtles will always be separated by an angle $\alpha = 120° = 2\pi/3$ radians and will always be equidistant from the origin, r. We utilize the positions and paths of A and B for our analysis; by symmetry, the behavior of C will be the same.

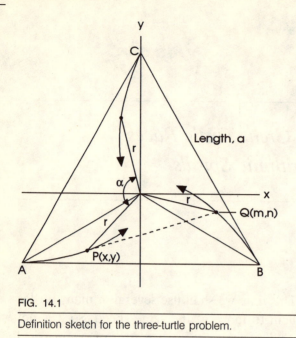

FIG. 14.1

Definition sketch for the three-turtle problem.

Expressing the coordinates (x, y) of point P and (m, n) of point Q in terms of the angles θ_A and θ_B, we establish the relationships

$$m = -\frac{1}{2}(x + \sqrt{3}\,y); \quad n = \frac{1}{2}(\sqrt{3}\,x - y). \qquad (14.1)$$

The criterion to "...always head directly toward..." is

$$\frac{dy}{dx} = \frac{n - y}{m - x}. \qquad (14.2)$$

Substituting (14.1) into (14.2) yields the differential equation

$$\frac{dy}{dx} = \frac{\sqrt{3}\,y - x}{\sqrt{3}\,x + y}. \qquad (14.3)$$

This equation is easily solved with the substitution $y = wx$. Carrying out some algebra and simple integrations, we obtain

$$\log_e \sqrt{x^2 + y^2} = -\sqrt{3} \arctan \frac{y}{x} + k, \qquad (14.4)$$

where k is an integration constant. This expression can be written in polar coordinate form by substituting $r = \sqrt{x^2 + y^2}$ and $\theta = \arctan(y/x)$:

$$\log_e r = -\sqrt{3}\,\theta + k.$$

To determine the value of the constant k, we employ the initial condition for either A, B, or C. Selecting the last, we have $r = a/\sqrt{3}$ when $\theta = 90° = \pi/2$ radians. This gives the answer:

$$\log_e \frac{\sqrt{3}\,r}{a} = \sqrt{3}\left(\frac{\pi}{2} - \theta\right), \qquad (14.5)$$

or in final form

$$r = \frac{a}{\sqrt{3}} e^{\sqrt{3}(\pi/2 - \theta)}. \qquad (14.6)$$

As we know, this is the equation of the logarithmic or equiangular spiral. In chapter 9, we learned that this beautiful curve makes its appearance in many places in nature. In hydrodynamics, for example, the logarithmic spiral is produced when a vortex flow is combined with a source or sink flow—like the spiral you get when you drain the bathtub. Numerous properties of this interesting curve are discussed by Lockwood (1961), by Steinhaus (1969), and by Thompson (1961).

Back to the three turtles. Their paths are shown in figure 14.2. Here's a good question: How far does each turtle go before it collides with the others at the center of the triangle? To answer this question we write the expression

$$(ds)^2 = (r\,d\theta)^2 + (dr)^2, \qquad (14.7)$$

where ds is an incremental length along the turtles' path and dr and $d\theta$ are incremental changes in r and θ. From this relation-

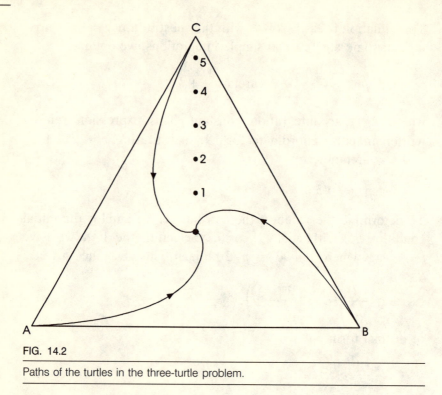

FIG. 14.2

Paths of the turtles in the three-turtle problem.

ship we obtain the expression

$$ds = \sqrt{r^2 + \left(\frac{dr}{d\theta}\right)^2}\, d\theta.$$ (14.8)

Using (14.6) in (14.8), we find the length of the travel path, $S = (2/3)a$. Thus, each turtle travels a distance equal to two-thirds of the length of the triangle's side. If $a = 1.0$ m $= 100$ cm and $v = 1.0$ cm/s, the turtles will meet at the center of the triangle in $T = S/v = 66.7$ s.

The obvious next problem concerns four turtles and a square. In this case, it is not difficult to show that the turtle paths are described by the equation

$$r = \frac{a}{\sqrt{2}} e^{\pi/4 - \theta},$$ (14.9)

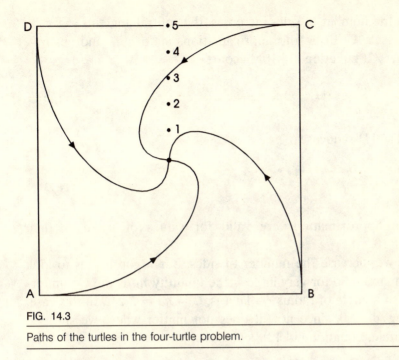

FIG. 14.3

Paths of the turtles in the four-turtle problem.

and the length of the path is $S = a$. Collision time is 100 seconds. This case is shown in figure 14.3.

Let us move on to five turtles and a pentagon, six turtles and a hexagon, eight turtles and an octagon, and n turtles and an n-gon. The general case solution is

$$r = a\left(\frac{\cos(\alpha/2)}{\sin \alpha}\right) e^{[(1 - \cos \alpha)/\sin \alpha](\theta_0 - \theta)}, \qquad (14.10)$$

in which $\alpha = 2\pi/n$ and n is the number of sides of the n-gon; θ_0 is a constant whose value is determined by the geometry of the initial condition.

The length of the path for the general case is

$$S = \left(\frac{1 + \cos \alpha}{\sin^2 \alpha}\right) a, \qquad (14.11)$$

where in each instance a is the length of a side of the n-gon.

As the number of sides increases, the central angle, $\alpha = 2\pi/n$, decreases. Utilizing the approximations $\sin \alpha = \alpha$ and $\cos \alpha = 1 - \alpha^2/2$, equation (14.10) becomes

$$r = \frac{na}{2\pi} e^{(\pi/n)(\theta_0 - \theta)}, \tag{14.12}$$

and (14.11) reduces to

$$S = \frac{n^2 a}{2\pi^2}. \tag{14.13}$$

These approximations are valid for values of n larger than about 10.

As we increase the number of sides, our n-gon begins to look like a circle. So for very large n the quantity na is the circumference of a circle of radius r_0. That is, $C = na = 2\pi r_0$. In this case, setting $\theta_0 = 0$, since now it does not matter where we start on the circle, equation (14.12) becomes

$$r = r_0 e^{-\pi\theta/n}, \tag{14.14}$$

and from (14.13) the total length of a path is

$$S = nr_0/\pi. \tag{14.15}$$

Are you ready for the great experiment? We applied for and received a government grant to carry out an experiment featuring 360 small turtles spaced at 1° intervals around the circumference of a circle of radius $r_0 = 10$ m. Each turtle has instructions to head directly toward the turtle immediately ahead and to move at a speed $v = 1.0$ cm/s.

Backing up briefly, we can establish that one of the properties of the logarithmic spiral is that it always intersects the radius vector r at a constant angle β. In our turtle problem we determine that this intersect angle $\beta = \alpha/2$, where $\alpha = 2\pi/n$ is the central angle. For example, if $n = 3$, then $\alpha = 2\pi/3$ and $\beta = \pi/3$ radians $= 60°$. If $n = 4$, then $\alpha = \pi/2$ and $\beta = \pi/4$ radians $= 45°$.

In our great experiment, $n = 360$, and so $\alpha = \pi/180$ and $\beta = \pi/360$ radians $= 0.5°$. Consequently, in our 360-gon, which is almost but not quite a circle, each turtle heads $0.5°$ inboard from the tangent to the equivalent circle.

Since we have a large value of n, equations (14.14) and (14.15) can be utilized. From (14.14) we see that $r = r_0 = 10$ m when $\theta = 0$. If $\theta = 2\pi$ radians $= 360°$ (i.e., one complete revolution of marching turtles around the circle), we determine that $r = 9.466$ m. Thus, the turtle circle has shrunk by about 53.4 centimeters after one revolution. In this first circuit they traveled approximately 61.6 m; since their speed is $v = 1.0$ cm/s, it took them 6,120 seconds or around 1 hour and 42 minutes for the first lap.

The second lap takes a bit less time since the circle is slightly smaller. The third lap is shorter still, and so on. From equation (14.15) we determine that the total distance traveled by each turtle to get to the center is $S = 1,146$ m and it takes 31 hours and 50 minutes to get there. We prefer to ignore what happens when 360 tired turtles all arrive at $r = 0$ at the same instant—a pretty nasty traffic pileup.

There is one feature in common in all of the preceding problems: the turtles are pursuing one another on *plane* surfaces. An interesting generalization of our turtle problem is given by Aravind (1994), in which the plane surface is replaced by a *spherical* surface (constant positive curvature) and by a *hyperbolic* (or pseudosphere) surface (constant negative curvature).

A final word regarding a recommended reference. Virtually all of the topics covered in our book are examined in a clear and concise way in the comprehensive volume by Gellert et al. (1977). If you have a serious interest in the study of mathematics, this excellent reference book should be in your personal library.

15

How Many People Have Ever Lived?

In 1990 the population of the world was approximately 5.32 billion people. This is an increase of 844 million over the 1980 population, which was an increase of 755 million over the 1970 population, which was an increase of 671 million over the 1960 population, . . . , and so on.

Interesting, but perhaps we are going the wrong way in time. Who cares about the population of the past—the demography of yesterday? We want to know about the population of the future—the shape of things to come. Two comments: First, the best way to make forecasts, at least for the near future, is to base projections on past and present information. Second, in the next chapter we shall indeed examine the topic of the world's population in the years to come.

So let us start by taking a look at the past. We begin with the following simple relationship:

$$\frac{dN}{dt} = aN, \tag{15.1}$$

in which N is the magnitude of a particular growing quantity at time t and a is a growth coefficient or an interest rate. This equation indicates that the *rate* at which a quantity is growing in magnitude is assumed to be directly proportional to the *magnitude* at that instant.

The solution to equation (15.1) is

$$N = N_0 e^{at}, \tag{15.2}$$

where e, of course, is the base of natural logarithms and N_0 is the magnitude of N at time $t = 0$. This is the equation for so-called exponential growth.

Without getting too complicated with our mathematics, (15.1) can be written in a more generalized form:

$$\frac{dN}{dt} = a(N)N, \tag{15.3}$$

which is the same as (15.1) except that we have replaced a with the quantity $a(N)$. This symbol says simply that the growth coefficient a is no longer constant but instead depends on the value of N. Let us be more specific about the form of $a(N)$. We shall say that the growth coefficient or interest rate is directly proportional to N; that is, $a(N) = aN/N_0$. This gives

$$\frac{dN}{dt} = \left(\frac{aN}{N_0} \right) N = \frac{a}{N_0} N^2. \tag{15.4}$$

The quantity N_0 is brought into the analysis at this point in order to keep the dimensions of the equation as simple as possible.

We note in (15.4) that the rate at which the particular quantity (e.g., population) is growing, dN/dt, is proportional not to the *first* power of N as in exponential growth, but to the *second* power, that is, the growth is like N squared. This kind of growth has been termed *coalition* growth by von Foerster et al. (1960).

As we shall see, coalition growth is truly "explosive" growth. It involves an increase to an infinite value of N in a finite time and it features a continuously shrinking "doubling time." It incorporates, for example, the unbelievably good deal you got at your local financial house: your probably insane banker agrees to make your savings interest rate directly proportional to the amount you have on deposit (e.g., if you have $1000 in your

TABLE 15.1

Population of the world, 1650 to 1990

Year	t years	N billion people
1650	0	0.510
1700	50	0.625
1750	100	0.710
1800	150	0.910
1850	200	1.130
1900	250	1.600
1950	300	2.565
1960	310	3.050
1970	320	3.721
1980	330	4.476
1990	340	5.320

account you get 4% interest; if you have $2000 you get 8%; $3000 gives 12%).

The solution to equation (15.4) is the amazingly simple expression

$$N = \frac{N_0}{1 - at},$$
(15.5)

where N_0 is the value of N when $t = 0$. This is the equation for coalition growth. Because of its mathematical form it is also called *hyperbolic* growth.

We go back to the population of the world. Information concerning the population is given in table 15.1 for the period from 1650 to 1990. We select 1650 as the year for which $t = 0$. The data shown in the table are plotted in figure 15.1. For the moment, disregard the solid curve shown in the figure.

It is observed that for a long time the world's population increased very slowly. Indeed, not until the beginning of the twentieth century did the population begin to rise sharply, and only after about 1950 did the population show really alarming increase.

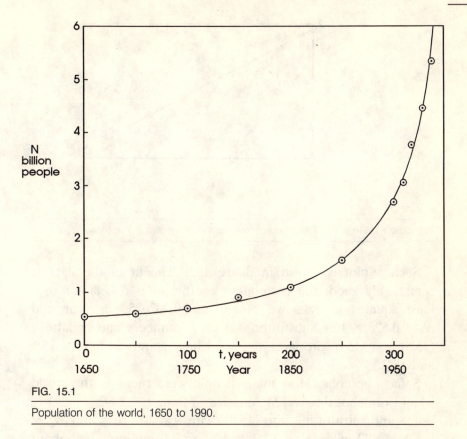

FIG. 15.1

Population of the world, 1650 to 1990.

We suspect that this growth of population may have been more rapid than exponential growth. Hence, for starters, we assume it may be hyperbolic (or coalition) growth. To confirm this we need to see how the data "fit" the mathematical model. It is easy to rewrite equation (15.5) in the form

$$\frac{1}{N} = \frac{1}{N_0} - \frac{a}{N_0}t. \tag{15.6}$$

In the language of analytic geometry, this equation has the linear form $y = k_0 + k_1 x$, where k_0 and k_1 are constants. Accordingly, if $1/N$ is plotted against t we should get a straight line if our assumption of hyperbolic growth is correct. The constants k_0 and k_1 provide the values of N_0 and a.

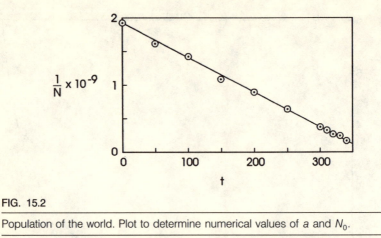

$\frac{1}{N} \times 10^{-9}$

t

FIG. 15.2

Population of the world. Plot to determine numerical values of a and N_0.

Such a plot is shown in figure 15.2. The fit of the data is remarkably good; the correlation coefficient is 0.9990. From a least squares analysis we obtain $a = 0.002675$ per year and $N_0 = 0.525 \times 10^9$. Substitution of these numbers into equations (15.5) and (15.6) produces the solid lines shown in figures 15.1 and 15.2.

Some questions: How many people were there in the world when Columbus discovered America? Taking $t = 1492 - 1650 = -158$ and substituting into (15.5) (with $a = 0.002675$ and $N_0 = 0.525 \times 10^9$) gives $N = 0.369 \times 10^9$ or 369 million. How about that very eventful year 1066? Answer: 205 million. What was the world's population in the year 4000 B.C.? From equation (15.5) we get 33 million.

It is risky business to make population projections very far into the future based on some kind of formula or equation. It is equally risky to estimate or try to calculate past populations. Even so, our almost pitifully simple equation (15.5) gives answers about previous populations that agree amazingly well with values obtained by anthropologists and demographers using entirely different methodologies.

For example, Deevey (1960) estimates that the world's population was 125,000 a million years ago; equation (15.5) gives 196,000. Westing (1981) cites a population of 3,000,000 in the year 40,000 B.C.; we get 4,670,000. Austin and Brewer (1971) indicate a population of 100 million for the year 0 A.D.; we compute 97 million; and so on.

Our main question is: how many people have ever lived on earth? In 1990 the world's population was 5.32 billion. Is this a large or small percentage of the total number who have ever lived?

This question is answered by returning to equation (15.5). We integrate this equation to determine the total number of person-years, M, of all who have ever lived. That is,

$$M = \int_{t_0}^{t} N\,dt = N_0 \int_{t_0}^{t} \frac{dt}{1 - at} = \frac{N_0}{a} \log_e \left(\frac{1 - at_0}{1 - at} \right), \qquad (15.7)$$

where we take $t_0 = -1,000,000$ as the date from which we start counting the number of people who ever lived.

For example, how many people lived during the period $t_0 = -1,000,000$ and $t = -500,000$? Substituting numbers into (15.7) gives $M = 135.6 \times 10^9$ person-years. Assuming a life span of duration $\bar{t} = 25$ yr gives $P = M/\bar{t} = 5.4 \times 10^9$ persons. In other words, about 5.4 billion people lived and died during the 500,000 years between 1,000,000 B.C. and 500,000 B.C.

Some anthropologists and historians indicate that the year 4000 B.C. marks the "dawn of civilization." If so, how many people lived in the predawn period, 1,000,000 B.C. to 4000 B.C.? From (15.7) we determine $M = 1,003 \times 10^9$ person-years for that period and, with $\bar{t} = 25$ yr as the average life span, obtain $P = 40.1$ billion people.

A listing of the cumulative number of people who have ever lived, commencing with the year 1,000,000 B.C., is shown in table 15.2. To follow the scheme of other studies of the subject, the average life span is taken as $\bar{t} = 25$ yr.

As they say, the bottom line is the answer. The last entry of table 15.2 indicates that through the year 1990, about 80 billion people have lived on earth. Some observations:

The world's population in 1990 was 5.32 billion. This is about 6.5% of the number of people who have ever lived. Said another way, about one person of fifteen who have ever lived is currently living.

TABLE 15.2

Cumulative number of people who have lived
since the year 1,000,000 B.C.

To (B.C.)	P billions	To (A.D.)	P billions	To (A.D.)	P billions
500,000	5.4	0	48.7	1800	66.0
250,000	10.8	1000	54.1	1900	70.6
100,000	17.9	1500	59.3	1950	74.7
10,000	34.7	1650	62.0	1980	78.8
4,000	40.1	1700	63.1	1990	80.8

About half of the 80 billion people lived between 1,000,000 B.C. and 4000 B.C.; the other half since then.

During the nearly 2,000-year period from the year 0 A.D. to 1990, about 32 billion people have lived. This is 40% of the total who have ever lived.

Our final answer, $P = 80$ billion, agrees fairly well with results obtained by others; they range from 50 billion to 110 billion. Suggested references are Deevey (1960), Goldberg (1983), Keyfitz (1966), and Westing (1981).

We conclude our chapter with a word about the doubling time. By definition, this is simply the time required for a growing quantity to double in magnitude. For example, a quantity may be following an exponential growth relationship. If so, from equation (15.2) it is easy to establish that the doubling time, t_2, is

$$t_2 = \frac{\log_e 2}{a} = \frac{0.693}{a} = \frac{70}{a(\%)}. \tag{15.8}$$

For example, if the population growth rate of a certain country is $a = 3.5\%$ per year, then the doubling time is 20 years. This type of exponential population growth is sometimes called Malthusian growth. We note that one of the features of such growth is a constant doubling time.

In contrast, suppose that a quantity is growing according to a hyperbolic growth relationship. From equation (15.5) we obtain the following expression for the doubling time:

$$t_2 = \frac{1 - at}{2a}. \qquad (15.9)$$

In this case, the doubling time, t_2, is not constant, as in exponential growth, but instead changes with time. In the year 1000 A.D., for example, $t = 1000 - 1650 = -650$, and so, from equation (15.9), the doubling time $t_2 = 512$ yr. In 1650, $t_2 = 187$ yr; in 1950, $t_2 = 37$ yr; in 1990, $t_2 = 17$ yr.

This is getting to be rather scary. Does this mean that in the year $1990 + 17 = 2007$, there will be $2 \times 5.52 = 10.64$ billion people in the world? And then it doubles again nine years after that? We shall examine these alarming questions in our next chapter, "The Great Explosion of 2023."

16

The Great Explosion of 2023

Our previous chapter concluded on a note depicting a very alarming situation: (1) a 1990 world population of over five billion people and a population doubling time of around 17 years and (2) a projection to a year-2000 population of over eight billion and a doubling time of around 12 years. Clearly, it is urgent that something be done about this serious crisis of rampant growth of world population.

However, we need to look at the data more closely. Before we try to resolve this incredible crisis, perhaps it is necessary to examine how serious it really is. So we return to the equation that describes so-called coalition or hyperbolic growth:

$$N = \frac{N_0}{1 - at},\tag{16.1}$$

where N is the population at time t and a is the growth coefficient; N_0 is the population at time $t = 0$. We obtained the values $a = 0.002675$ per year and $N_0 = 0.525 \times 10^9$ based on $t = 0$ in the year 1650.

Some populations are listed in table 16.1, computed from equation (16.1), for various years commencing with 1650 and ending with 2023. Magnitudes of observed populations are also shown and the differences between computed and observed pop-

TABLE 16.1

Population of the world, N, in billions

Year	t	N_{obs}	N_{comp}	$N_{comp} - N_{obs}$
1650	0	0.510	0.525	+0.015
1700	50	0.625	0.606	−0.019
1750	100	0.710	0.717	+0.007
1800	150	0.910	0.877	−0.033
1850	200	1.130	1.130	0.000
1900	250	1.600	1.586	−0.014
1950	300	2.565	2.660	+0.095
1960	310	3.050	3.077	+0.027
1970	320	3.721	3.649	−0.072
1980	330	4.476	4.482	+0.006
1990	340	5.320	5.808	+0.488
2000	350		8.248	
2010	360		14.225	
2020	370		51.650	
2022	372		109.016	
2023	373		245.161	

ulations are indicated. Several important facts can be seen in the table:

For the years 1650 through 1980 there is surprisingly close agreement between computed and observed populations. This is apparent in the right-hand column of the table.

For 1990, however, there is a difference of +0.488. That is, the computed population is nearly 500 million more than the observed. This is a very interesting and important point. We shall come back to it later.

For the years 2000 and beyond, computed populations rise to ridiculous values: nearly 52 billion in 2020 and over 245 billion on, say, January 1, 2023.

Okay, here it comes. Doomsday is established by setting the denominator of equation (16.1) equal to zero. This gives $t_e = 1/a = 373.832$, where t_e stands for explosion time. Since $t = 0$ corresponds to 1650, then explosion year is 2023.832.

So 2023 will be the year of the great explosion. Both Keyfitz (1968) and von Foerster et al. (1960) arrive at about the same conclusion. Specifically, on November 1, 2023 there will be an infinite number of people in the world and the doubling time will have shrunk to zero.

Interesting. Now back to reality. Of course, the world's population is going to continue to increase in the years to come. However, obviously the population will not become infinite. Inevitably and increasingly there will be forces that will slow down, terminate, or even reverse the growth rate.

These forces imposing limitations to population growth are comprised of agricultural, demographic, economic, environmental, scientific, sociological, and technological components. Collectively, they specify and emphasize the finite capacity of the world for maintenance of a human population.

Apparently, population crowding effects began to appear, on a worldwide basis, during the decade of the 1980s. We noted the very important fact that the computed population for 1990 was almost 500 million more than the observed population. It is remarkable that after a million years of human population increases, crowding effects and limitations-to-growth factors made their first appearances only quite recently.

Long ago it was found that an appropriate way to slow down exponential or Malthusian types of growth phenomena was to attach a crowding or finite-resources term to the growth equation. This turned a skyrocketing exponential growth into a stabilized logistic growth. It seems logical to modify our hyperbolic growth equation in a similar fashion.

To follow the approach of Austin and Brewer (1971), we arbitrarily attach a growth limitation term to the equation for unrestrained explosive hyperbolic growth. That is,

$$\frac{dN}{dt} = \frac{a}{N_0} N^2 \left(1 - \frac{N}{N_*} \right), \tag{16.2}$$

where the quantity in parentheses requires that the growth rate, dN/dt, become zero when $N = N_*$. This quantity, N_*, is the carrying capacity, that is, the maximum population the earth is able to sustain. The solution to equation (16.2) is

$$at = \left(1 - \frac{N_0}{N}\right) + \frac{N_0}{N_*} \log_e\left(\frac{N_* - N_0}{N_0}\right)\left(\frac{N}{N_* - N}\right),$$

$$(16.3)$$

which we will call the modified coalition growth equation. This expression is not as formidable as it might appear; indeed, it is quite easy to handle. We note that if N_* approaches infinity, that is, resources are sufficiently large to support an infinite population, equation (16.3) reduces to (16.1) as expected.

Unfortunately, (16.3) is in the form that says "t is a function of N" and not "N is a function of t," as we would prefer. This is not a problem; we still know how the world's population N increases with time t.

At this point we shift the time origin, $t = 0$, from the year 1650 to a more recent date: 1980. Also, with computations involving populations and growth rates through 1990, a value of N_* is obtained. These calculations provide the following: $a = 0.0303$, $N_0 = 4.450 \times 10^9$, and $N_* = 10.0 \times 10^9$ with $t = 0$ corresponding to 1980.

Populations calculated from equation (16.3) are listed in table 16.2. We compute a 1995 world population of about 5.73 billion and a year 2000 population of around 6.22 billion. These values are in good agreement with those given by the United Nations (1993).

A plot of equation (16.3) is shown in figure 16.1, in which we note several things:

In the years to come, the world's population certainly continues to increase. However, because of the stabilizing effect of the crowding term in (16.2), the explosive increase of uncontrolled hyperbolic growth is avoided.

As shown in figure 16.1, the rate of population increase reaches a maximum value in 2004 when the population is 6.667 billion people. The corresponding annual increase is 101 million people.

TABLE 16.2

Population of the world, *N*, in billions
Computed from modified coalition growth equation.

N	Year
3.0	1954.8
4.0	1973.6
5.0	1986.9
6.0	1997.7
7.0	2007.7
8.0	2018.2
9.0	2032.2
9.5	2044.0
9.9	2068.9

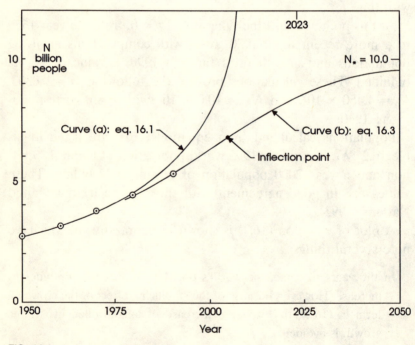

FIG. 16.1

Population of the world with projections leading to (a) infinite population in 2023 and (b) stabilized population of $N_* = 10.0$ billion.

This maximum occurs when the growth curve passes through the so-called inflection point, as shown in the figure. After that, the rate begins to decline and eventually becomes zero.

One thing needs to be emphasized. Projections of future populations are not made on the basis of simply fitting curves to observed data. Demographic analyses of trends of the world's population are done on a country-to-country basis and take into account such things as age distribution, life expectancy, migration, mortality, and fertility of a particular nation's population. World trends are established as the aggregation of country and regional trends. For those interested in these topics, a good place to start is the interesting article by Coale (1974). More advanced treatments of the subject are provided by Keyfitz (1968), by Pollard (1973), and by Song and Yu (1988).

Complementing precise demographic analyses are mathematical frameworks such as our modified coalition growth model. These models can answer a lot of questions.

PROBLEM 1.   The growth rate, dN/dt, is a maximum when its derivative, d^2N/dt^2, is equal to zero. Utilizing equations (16.2) and (16.3), show that the coordinates and slope of the growth curve at the inflection point are

$$N_i = \frac{2}{3}N \; ; \; \left(\frac{dN}{dt}\right)_i = \frac{4aN_*^2}{27N_0} \; ;$$

$$at_i = \left(1 - \frac{3N_0}{2N_*}\right) + \frac{N_0}{N_*} \log_e 2\left(\frac{N_*}{N_0} - 1\right). \tag{16.4}$$

PROBLEM 2.   In the previous chapter we established that during the period from 1,000,000 B.C. to 1950, the number of person-years, M, was about 1,867 billion. By exact, numerical, or graphical integration of equation (16.3), confirm that the value of M between 1950 and 2050 will be about 625 billion.

We have taken a fairly good look at the world's population. Here is a point on which to conclude. From equation (16.2) we determine that during 1995 the world's population increased by about 95 million people. If these 95 million were to comprise an entirely new country, where would it rank, in 1995 population, among the world's nations? The answer is number 11, after China, India, United States, Indonesia, Brazil, Russia, Pakistan, Japan, Bangladesh, and Nigeria. Just think: an enormous new country every year!

17

How to Make Fairly Nice Valentines

Have you noticed how dramatically the cost of greeting cards has risen during recent decades? Do you realize that each year we pay trillions of dollars for birthday cards and holiday cards? Many people find this difficult to believe.

Well, we can begin reducing our budgets for greeting cards by creating our own, at least some of them. The problem is that unless one has considerable artistic talent, it is impossible to trim expenses by making one's own fantastically beautiful Christmas cards, Labor Day cards, and Groundhog Day cards. Likewise, homemade Easter cards and Thanksgiving cards would probably look pathetic. By the same token, it is futile to try to do much about constructing your own birthday cards. What could really be done beyond slapping on some important anniversary numbers like 5, 18, 39, and 65?

However—here comes the good news—as mathematicians we can save substantial sums of money by fabricating our own greeting cards for those holidays with which simple mathematical shapes are associated. For example, for Halloween there is the symbol of the pumpkin, which is something like an oblate ellipsoid whose equation is $x^2/a^2 + y^2/b^2 + z^2/c^2 = 1$. We could draw ellipses on stiff paper, paint them orange, and mail them out by the dozens.

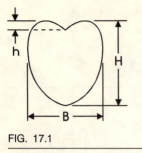

FIG. 17.1

Definition sketch for the valentine.

Or what about Saint Patrick's Day and the symbol of the shamrock? This is a three-leaf clover and looks like a trifolium, whose equation is $r = \cos\theta(4a\sin^2\theta - b)$. We could compute the curve, plot it on many pieces of cardboard, color them green, and send out as many as we like.

Best of all, of course, would be Valentine's Day. This is the big one! We could bring about industrial strength savings of cash by manufacturing our own valentines.

As we shall see shortly, there are numerous mathematical equations we could employ to construct valentines—some easy and some not. Furthermore, unlike pumpkins and shamrocks, a self-made mathematically perfect valentine to your loved one could result in substantial rewards—on a par with those ensuing from gifts of red roses and chocolates.

An Easy Valentine: A Circle and a Parabola

Before we start the mathematical construction of valentines, it is necessary to decide the overall proportions and dimensions we want. A definition sketch is shown in figure 17.1.

Arbitrarily, we select the proportion $B/H = 0.8$. If you prefer a square valentine, then take $B/H = 1.0$. For a fat valentine, take $B/H = 1.5$, and for a skinny one use $B/H = 0.5$. We will not specify the "plunge" distance, h; this dimension will be an outcome of the mathematics. Finally, for our computations we shall use $H = 10$ cm and $B = 8$ cm. This will be a nice fit for a small envelope.

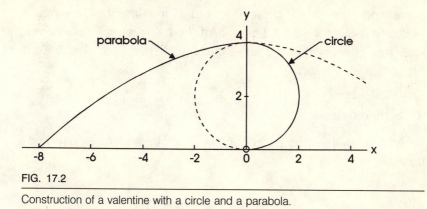

FIG. 17.2

Construction of a valentine with a circle and a parabola.

Our first, and mathematically easiest, valentine will be composed of half of a circle and half of a parabola. In figure 17.2, the valentine has been cut in half and laid on its side. The abscissa, x, coincides with the axis of the valentine and the ordinate, y, is perpendicular to x. The origin of this Cartesian or rectangular coordinate system is shown in the figure.

The Circle

We start with the circle. Since $B = 8$ cm then $B/2 = 4$ cm and so the radius of the circle must be $r_0 = 2$ cm.

Next, we need to call on analytic geometry. The equation of a circle of radius r_0 centered at $x = x_0$ and $y = y_0$ is

$$(x - x_0)^2 + (y - y_0)^2 = r_0^2. \tag{17.1}$$

In our problem, $x_0 = 0$, $y_0 = 2$, and $r_0 = 2$. Substituting these numbers into (17.1) and solving for y gives

$$y = 2 \pm \sqrt{4 - x^2}. \tag{17.2}$$

We now substitute values of x into this equation and compute the corresponding values of y. For example, if $x = +1$ or -1, we get $y = 2 + \sqrt{3}$ and $2 - \sqrt{3}$.

Computations like this provide the (x, y) coordinates of points of the circle. Through a number of such points we pass a smooth curve, and there is our circle. (Of course, for this curve you might prefer to use a compass.) The half of the circle we will use for our valentine is shown as the solid curve in figure 17.2; the unused other half is dashed.

The Parabola

Next we analyze the parabolic portion. The general equation of a parabola is

$$y = k_0 + k_1 x + k_2 x^2, \tag{17.3}$$

where k_0, k_1, and k_2 are constants. Now the height of our valentine is $H = 10$ cm and the top 2 cm consists of the semicircle; therefore the bottom parabolic portion must be 8 cm. So, with reference to figure 17.2, we need a parabola that passes through the points $x = -8$, $y = 0$ (and, by symmetry, $x = +8$, $y = 0$), and also $x = 0$, $y = 4$. Using equation (17.3), a bit of algebra yields $k_0 = 4$, $k_1 = 0$, and $k_2 = -1/16$. So the equation of the parabola is

$$y = 4 - \frac{1}{16}x^2. \tag{17.4}$$

From this expression, we plot the parabola. For example, if $x = \pm 4$ cm, (17.4) gives $y = 3$ cm.

The portion of the parabola we need for our valentine is shown as the solid curve in figure 17.2; part of the unneeded portion is dashed. The plunge distance for this valentine is $h = 2$ cm.

Only one more step is needed to complete this first design. We simply connect a mirror image to our semivalentine and we are ready for the red paint.

How much red paint do we need? This is a lowbrow way of asking the question: what is the area of our semivalentine? Well, the area of the semicircle is $A_c = (1/2)\pi r_0^2$. So, with $r_0 = 2$ cm, $A_c = 2\pi = 6.28$ cm^2.

What is the area of the semiparabola? It is fairly easy to show that the area of a parabola is equal to two-thirds of the area of the circumscribing rectangle. Accordingly, we get $A_p = (2/3)(8)(4) = 21.33$ cm^2. Consequently, the area of the semi-valentine is $A_{cp} = 27.61$ cm^2; the area of the entire valentine is double this amount, that is, $A = 55.22$ cm^2. Knowing how many valentines you need to make—that is, the number of sweethearts you have—you can now easily compute how much red paint, stiff paper, and envelopes you must buy.

PROBLEM   Now suppose you want to put some fancy tassel or ribbon along the edge of your valentine. To do this, you need to know the length of the perimeter of the valentine.

For the semivalentine, the length of the semicircular portion is easy to calculate; it is simply $S_c = (1/2)2\pi r_0 = 6.28$ cm. However, the length of the semiparabola is a bit more difficult to determine. Using the methods of elementary calculus, it is possible to show that the length of the parabola can be computed from the equation

$$S = \int_{x_1}^{x_2} \sqrt{1 + \left(\frac{dy}{dx}\right)^2} \, dx, \tag{17.5}$$

which in our problem becomes, using (17.1),

$$S_p = \int_{-8}^{0} \sqrt{1 + \frac{1}{64}x^2} \, dx = 9.18 \text{ cm}. \tag{17.6}$$

Consequently, the entire length of the perimeter of the valentine is $S = 30.92$ cm.

Perhaps this seems like a rather trivial problem. However, structural engineers face a similar computation involving the parabola when computing the length of the cables of large suspension bridges.

A More Difficult Valentine: A Logarithmic Spiral

In dealing with entirely different topics in other chapters, we ran across the famous mathematical curve called the logarithmic spiral. This curve, also known as the equiangular spiral, has been around for a long time. It was first studied in 1638 by the French mathematician René Descartes (1594–1650).

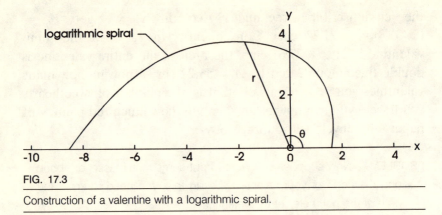

FIG. 17.3

Construction of a valentine with a logarithmic spiral.

The logarithmic spiral describes the shape of the beautiful Nautilus sea shell and the pattern formation in many kinds of plants and flowers. It is closely related to the remarkable golden number, $\phi = \frac{1}{2}(1 + \sqrt{5})$, and to the famous Fibonacci sequence, $1, 1, 2, 3, 5, 8, \ldots$

The equation of the logarithmic spiral is

$$r = r_0 e^{a\theta}, \tag{17.7}$$

in which r_0 and a are constants. This equation is plotted in figure 17.3.

In our first problem, involving the circle and the parabola, the analysis was carried out in a *rectangular* (x, y) coordinate system. We have now moved, with equation (17.7) to a *polar* (r, θ) coordinate system. Incidentally, we can easily shift from one system to the other because $x = r \cos \theta$ and $y = r \sin \theta$.

In these expressions, r is the distance from the origin, $(0, 0)$, to any point on the curve and θ is the angle between the x-axis and the r-line; r_0 is the value of r when $\theta = 0$. A reminder: When using (17.7), be sure to express θ in radians, not degrees, $(1 \text{ radian} = 360°/2\pi = 57.296°.)$

Half of our logarithmic spiral valentine is shown in figure 17.3. In order to make the overall dimensions of this valentine the same as the first one, that is, $H = 10$ cm and $B = 8$ cm, it is necessary to do some curve fitting. This operation is greatly simplified if we use some calculus.

PROBLEM 1.   If the equation of a plane curve is expressed in polar coordinates, $r = r(\theta)$, then the *slope* of the curve at any point (r, θ) is given by

$$\frac{dy}{dx} = \frac{(dr/d\theta)\sin\theta + r\cos\theta}{(dr/d\theta)\cos\theta - r\sin\theta}. \qquad (17.8)$$

Using (17.7) in (17.8), show that y is a maximum when $\tan\theta = -(1/a)$ and that x is a maximum when $\tan\theta = a$. This information enables you to determine that the values $r_0 = 1.432$ cm and $a = 0.561$ will assure that $H = 10$ cm and $B = 8$ cm. The plunge distance $h = 0.22$ cm.

   The total area of the valentine is $A = 60.25$ cm^2 and the entire perimeter is $S = 28.26$ cm.

PROBLEM 2.   Confirm these numerical values for area and perimeter. You can calculate the area from the equation

$$A = \frac{1}{2}\int_{\theta_1}^{\theta_2} r^2\, d\theta, \qquad (17.9)$$

and determine the arc length from the expression

$$S = \int_{\theta_1}^{\theta_2}\sqrt{r^2 + \left(\frac{dr}{d\theta}\right)^2}\, d\theta. \qquad (17.10)$$

A Much More Difficult Valentine: A Cardioid and a Lemniscate

   The analysis and design of this valentine are similar to our first one in which we connected a circle to a parabola. This time we are going to connect a cardioid to a lemniscate.

The Cardioid

   The word "cardioid" means a "heart-shaped curve," so we are off to a good start. Have you always wanted to know how to construct a cardioid but were too afraid to ask? The answer: Put two large coins of the same diameter on the table. Hold one of them down firmly. Then roll the other coin around the boundary

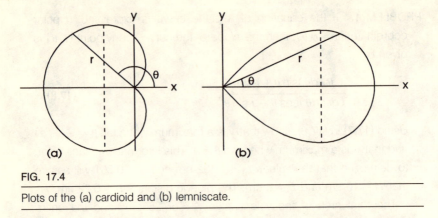

FIG. 17.4

Plots of the (a) cardioid and (b) lemniscate.

of the first without slipping. The original contact point between the two coins will trace a cardioid.

The equation of the cardioid, in a polar coordinate system, is

$$r = 2R(1 - \cos \theta),\qquad\qquad(17.11)$$

in which R is the radius of the coin, if you like. The entire cardioid is shown in figure 17.4(a). Now this may look like a heart but it sure doesn't look like a valentine. So we shall use only the right-hand portion and, in a moment, use something else for the left-hand portion. Figure 17.5 shows the design diagram for this one.

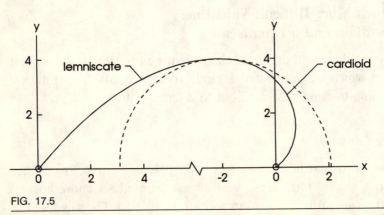

FIG. 17.5

Construction of a valentine with a cardioid and a lemniscate. The origin at the left refers to the lemniscate and that at the right to the cardioid.

As before, it is necessary that our cardioid-type valentine have the same overall dimensions: $H = 10$ cm and $B = 8$ cm. This requires that we compute the value of R in equation (17.11). To accomplish this, (17.8) is useful to determine the locations of the maximum x and maximum y points and the corresponding angles θ.

It turns out that the value $R = 1.538$ cm will force the cardioid through the point $r = 4.62$, $\theta = 120°$ or, equivalently, through $x = -2.31$, $y = 4.00$. Note that $y = 4$ is half the width of our valentine. We also determine that the plunge is $h = 0.77$ cm.

The portion of the cardioid we shall use is the solid line in figure 17.5; the unused portion is dashed.

The Lemniscate

Just as we attached a parabola to a circle to fabricate our first valentine, we are now going to attach a curve called a lemniscate to the cardioid to construct this one. The equation of the lemniscate is

$$r = L\sqrt{\cos 2\theta}. \tag{17.12}$$

The entire lemniscate is shown in figure 17.4(b); only the left-hand portion will be used. To simplify the algebra, we have selected a new origin for this part of the design.

The lemniscate has somewhat the shape of a leaf or a pointed circle. It looks like the trajectory of a high pop-fly to center field. If the curve is rotated about the x-axis, the resulting shape looks like a teardrop or—upside down—a hot-air balloon. In 1694, the noted Swiss mathematician Jakob Bernoulli (1654–1705) published a paper about this interesting curve. The lemniscate is a special case of another curve called the Cassinian ovals.

As before, we need to determine the value of L in equation (17.12) to assure that the final valentine will have the dimensions $H = 10$ cm and $B = 8$ cm. Again, (17.8) is helpful in determining the location of the maximum point of the lemniscate. We then match this point with the maximum point of the cardioid. The outcome is that $L = 11.30$ cm.

So, for our cardioid-lemniscate valentine, we have $R = 1.54$ cm and $L = 11.30$ cm. If you would like a nice problem to work on during your spare time, you can confirm that indeed $H = 10$ cm and $B = 8$ cm for this valentine.

You can also confirm, using equation (17.9), that the area of the cardioid portion of the entire valentine is $A_c = 20.52$ cm^2 and the area of the lemniscate portion is $A_l = 36.24$ cm^2. The total area is $A = 56.76$ cm^2.

Finally, utilizing (17.10), you might want to show that the perimeter of the cardioid portion of the entire valentine is $S_c = 12.31$ cm. A bit more difficult task is to establish that the perimeter of the lemniscate portion is $S_l = 16.44$ cm. The total perimeter is $S = 28.75$ cm.

Some Closing Comments

That should be sufficient, unless, of course, you want to create and analyze some other valentine configurations.

There are many references that deal with the subject of mathematical curves. Three of the best are Lawrence (1972), Lockwood (1961), and von Seggern (1990).

Finally, to help you prepare for next Valentine's Day, it is suggested that you plot, on a sheet of graph paper, the three valentines we analyzed in the preceding sections. Following this, you have two major decisions to make: (1) which valentine you prefer and (2) whether or not to go into mass production. Whatever you decide, it is recommended that you do not wait until next February to carry out this assignment.

18

Somewhere Over the Rainbow

And to Continue the Song

> ...bluebirds fly
> Birds fly over the rainbow
> Why then, oh why, can't I?

Nature is very generous. Frequently, and always gratuitously, it provides all of us with spectacular exhibitions of magnificent beauty. And surely, at or near the top of everyone's list of incredibly beautiful displays of nature would be the rainbow. For thousands of years, poets and artists, scientists and mathematicians have attempted to describe the rainbow with words and paintings, symbols and equations.

According to historical records, the noted Greek philosopher Aristotle (384–322 B.C.) made extensive studies of rainbows, trying to understand what they are and how they are created. But for many hundreds of years, the magnificent arc of color eluded mankind's efforts at quantitative description and mathematical analysis. During the long period of the Middle Ages, only the name Theodoric of Freiberg (c. 1310) stands out in connection with rainbows. The advances made by this Franco-German monk brought "rainbow theory" remarkably close to present-day points of view.

In the three hundred years following Theodoric, little progress was made by scientists and mathematicians in the analysis of rainbows. Then it all came together, as they say, in the seventeenth century. During that period, a sequence of very remarkable people developed and utilized the mathematics and physics needed to explain the phenomenon. Here is a list of these people:

Johannes Kepler, German, 1571–1630

Willebrord Snell, Dutch, 1591–1626

René Descartes, French, 1596–1650

Christiaan Huygens, Dutch, 1629–1695

Isaac Newton, English, 1643–1727

The advances made by these scientists, culminating in the work of Descartes and Newton, were so complete and accurate that we still refer to the theory of the rainbow as the Descartes–Newtonian theory. Indeed, their relatively simple geometrical optics solution to the problem explains all the main features of the rainbow; only a few minor aspects of rainbows require more complicated mathematics.

Rainbows Are in Many Places

An extremely pleasant surprise sometimes occurs when we suddenly look up or go around a corner and there is a magnificent rainbow. Even so, it is helpful to know where to look for rainbows and when to expect them. It is very simple: A bright sun needs to be at your back as you look toward a rain shower. Midmornings or earlier and midafternoons or later are best in order to catch low sun angles; this puts the bow higher in the sky.

Sometimes there are two rainbows. If so, the inner and more intense bow is called the primary rainbow. As we shall see, this bow is the arc of a circle whose radius is about 42°. Violet is on the inside of the bow and red is on the outside; all the other colors—blue, green, yellow, orange—are in between.

If we are lucky, we are able to see also the outer and less intense bow called the secondary bow. It has an angular radius of about 50°. The colors are reversed in the secondary bow; red is on the inside and violet on the outside.

However, we do not require rain showers to see rainbows. As long as we have a bright light at our back and a source of water drops in front, we can see them. For example, they appear in the spray at the base of a waterfall or near the jet of a garden hose or in the bow wave of a ship or motorboat. If you happen to be the pilot of an airliner with the sun behind you and a rainstorm ahead, maybe you will see the complete circle of the rainbow!

Some References about Rainbows

Over the years a great deal has been written about rainbows. An interesting book by Boyer (1987) gives a comprehensive history of the subject and an easy-to-understand presentation of the Decartes–Newtonian analysis of rainbows. A book by Greenler (1980) offers nontechnical descriptions and explanations about rainbows and includes a remarkable collection of color photographs.

An excellent survey article about the theory of rainbows is given by Nussenzveig (1977). The author includes an introduction to the more advanced topics of rainbow theory. Finally, Austin and Dunning (1988) and Minnaert (1993) are references that deal with the elementary mathematics and physics of rainbows.

Geometrical Analysis of Rainbows

Our study of rainbows begins with an analysis of the geometry involved in the phenomenon. For this, we use the principles of a branch of physics called *geometrical optics*. We start with figure 18.1. A ray of light from some source—for example, the sun—strikes the surface of a drop of water at point A. This ray is called the incident ray. As shown in the figure, the line A-A', perpendicular to the surface, defines the angle i of the incident ray. Another ray, named the reflected ray, leaves the surface of

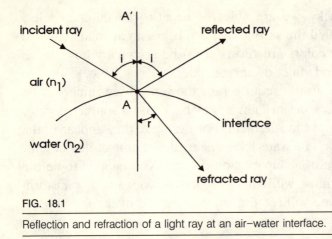

FIG. 18.1

Reflection and refraction of a light ray at an air–water interface.

the water drop at the same angle i. Thus, the angle of incidence is equal to the angle of reflection. We shall use this simple and intuitively obvious relationship throughout our analysis.

However, not all of the light of the incident ray is reflected back into the air in the reflected ray. Some portion of the light is refracted into the water at an angle r, as shown in the figure. This angle is called the angle of refraction. In 1618, the Dutch scientist Willebrord Snell (1591–1626) gave the following equation relating the angle of incidence i to the angle of refraction r:

$$n_1 \sin i = n_2 \sin r, \tag{18.1}$$

where n_1 and n_2 are the so-called refractive indices of air and water, respectively. The relationship of equation (18.1) is usually called Snell's law.

The numerical value of the refractive index depends on the type of medium through which the light ray passes. In a vacuum and in air, $n = 1$ and in water, $n = 1.33$. We simplify (18.1) by writing it in the form

$$\sin i = n \sin r, \tag{18.2}$$

with the understanding that we are now dealing with an air-water boundary. For example, if $i = 60°$ and $n = 1.33$, then (18.2) gives $r = 40.6°$. We note in figure 18.1 that the reflected ray is bent *toward* the perpendicular *A-A'*.

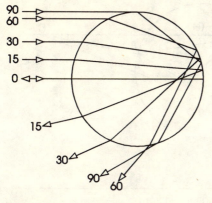

FIG. 18.2

The primary bow. Two refractions and one reflection of incident rays in a water drop, and the corresponding emergent rays.

The Primary Rainbow

First, we examine what is called the primary rainbow. As we shall see, the primary rainbow involves two refractions and one reflection of a light ray inside a raindrop. As shown in figure 18.2, light rays from the sun strike a spherical raindrop at various distances at or above a line passing through the center of the drop. The angle of incidence ranges from $i = 0°$ to $i = 90°$. At each impact point, part of the light of the ray is reflected back into the air and part is refracted into the water. For the moment, we are interested in only the latter.

As indicated in the figure, the refracted ray proceeds across the raindrop until it strikes the opposite boundary. Here a portion of the light is refracted into the air. More importantly, the remaining portion is reflected in the drop and travels to the front boundary of the drop. Once again, part of the light is reflected to another point on the inside of the drop. The remaining part is refracted into the air.

We take a closer look at the geometrical optics in figure 18.3. An incident ray strikes the drop at point A at some angle of incidence, i. Here, the ray is refracted to angle r and proceeds to

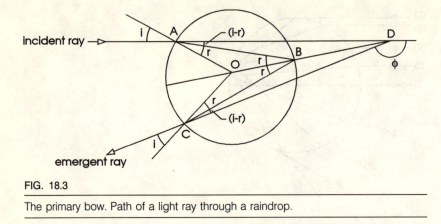

FIG. 18.3

The primary bow. Path of a light ray through a raindrop.

point B. The ray is then reflected to point C, where it is refracted into the air. We call this the emergent ray.

In figure 18.3, extensions of the incident and emergent rays define a point D and an angle ϕ. From the geometry of the figure, it is not hard to show that ϕ expresses the total change of angle between the incident ray and the emergent ray. Its value is

$$\phi = 2(i - r) + (180 - 2r). \tag{18.3}$$

A plot of this equation is shown in figure 18.4.

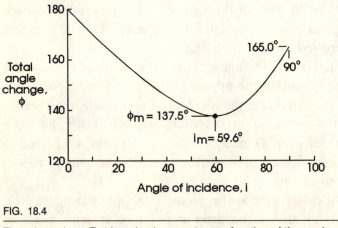

FIG. 18.4

The primary bow. Total angle change, ϕ, as a function of the angle of incidence, i.

Next, we ask the question, is there a value of the incidence angle i that gives a minimum value of total angle change, ϕ? To answer the question, it is necessary to use a bit of differential calculus. We rewrite equation (18.3) in the form

$$\phi = 180 + 2i - 4\arcsin\left(\frac{1}{n}\sin i\right), \tag{18.4}$$

in which we have used (18.2) to relate i and r.

To determine the minimum value of ϕ, equation (18.4) is differentiated with respect to i and the result is equated to zero. This gives the following answer:

$$\cos i_m = \sqrt{\frac{n^2 - 1}{3}}, \tag{18.5}$$

where i_m is the value of the incidence angle giving minimum ϕ. If $n = 1.33$, then $i_m = 59.6°$. Substituting this result into (18.4) yields $\phi_m = 137.5°$. These numerical values are identified by the solid dot on the curve in figure 18.4. A ray of light passing through a raindrop with these particular values of incidence and emergence angles is sometimes called the *Descartes ray*. A note in passing: In 1637, the French scientist René Descartes (1596–1650) was the first person to work out the geometrical optics of rainbows correctly.

We note from figure 18.4 that no light rays have a total angle change ϕ less than about 138°. So it is apparent that all rays emerging from the raindrop are confined to a circular cone with half-angle of $180° - 138° = 42°$. The apex of this cone is the eye of the observer and the axis of the cone is in the direction opposite to the sun.

Furthermore, since the curve displayed in figure 18.4 is quite flat near the minimum point, most of the emergent rays are at or near the 42° boundary of the cone. In this regard, if you would like another problem in calculus, you can show that the *average* value of ϕ, over the entire range of i, is $\overline{\phi} = 151.5°$. This corresponds to a cone half-angle of 28.5°.

Why Do Rainbows Have Colors?

So far so good. Perhaps we have now convinced ourselves that incident sunlight, after two refractions and one reflection within a spherical raindrop, produces emergent rays largely confined to a cone with 42° half-angle. Fine. But where does the color come from?

To answer this question we turn to results obtained by that remarkable English scientist, Isaac Newton (1643–1727). Along with the large number of other incredible advances he made, Newton was able to show, in experiments carried out in 1666, that a ray of sunlight is composed of an entire spectrum of colors. Each color of the ray has a particular numerical value of the refractive index n as the ray passes through a transparent medium such as water.

For example, Newton showed that in water, red has a refractive index $n = 1.332$, and at the other end of the spectrum, violet has a refractive index $n = 1.344$. The refractive indices of all the other colors—orange, yellow, green, blue—have numerical values between those of the red and violet extremes.

For our purpose, the rest is easy. Starting with red, we substitute $n = 1.332$ into equation (18.5) to compute i_m and then use (18.4) to determine that $\phi_m(\text{red}) = 137.8°$. This corresponds to a cone half-angle of 42.2°. At the other end of the spectrum, violet with $n = 1.344$ gives $\phi_m(\text{violet}) = 139.5°$, which provides a cone half-angle of 40.5°.

A quick summary of the results to here: In the primary rainbow, two refractions and one reflection of sunlight rays within a raindrop produce a bow with a cone half-angle of about 42°. In other words, the rainbow is the arc of a circle with 42° angular radius. The red color in sunlight yields a cone half-angle of 42.2° and the violet color gives a half-angle of 40.5°. The slight difference between these two angles provides the spectrum of color in a rainbow. Since the cone angle for red is larger than the cone angle for violet, it is clear that red is on the outside of the primary rainbow and violet is on the inside. The other colors are in between.

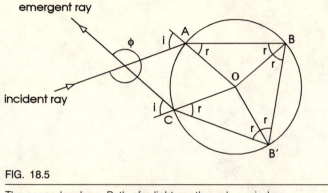

FIG. 18.5

The secondary bow. Path of a light ray through a raindrop.

The Secondary Rainbow

Our analysis of the so-called secondary rainbow is identical to that of the primary rainbow with one exception: this time there are two reflections, instead of one, of a ray of sunlight within the raindrop. So, as shown in figure 18.5, for the secondary bow a ray of light receives two refractions and two reflections inside the drop.

In this case, application of geometrical optics leads to the expression

$$\phi = 2(i - r) + 2(180 - 2r), \tag{18.6}$$

in which ϕ is the total change in angle between the incident ray and the emergent ray. Note that this equation is slightly different from equation (18.3), which gives the total angle change for the primary rainbow.

A plot of (18.6) is shown in figure 18.6. It is apparent that again there is a minimum value of ϕ. Substituting equation (18.2) into (18.6) and then differentiating with respect to i yields the result

$$\cos i_m = \sqrt{\frac{n^2 - 1}{8}}. \tag{18.7}$$

Again, the refractive index of sunlight in water is $n = 1.33$. Substituting this value into (18.7) gives $i_m = 71.9°$. Using this

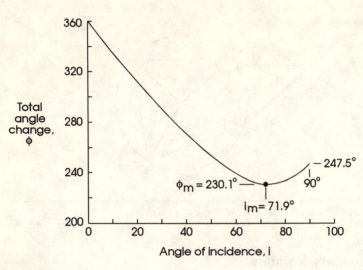

FIG. 18.6

The secondary bow. Total angle change, ϕ, as a function of the angle of incidence, i.

answer in (18.6) yields the minimum value $\phi_m = 230.1°$. So, in the secondary rainbow, most of the rays are confined to a circular cone with half-angle $230° - 180° = 50°$.

As we did before, using $n = 1.332$ (red) and $n = 1.344$ (violet), we easily determine that red appears at a cone half-angle of $50.6°$ and violet at $53.7°$. Clearly, in the secondary rainbow the colors are reversed from the primary rainbow. Violet is now on the outside of the bow and red is on the inside; all the other colors—orange, yellow, green, and blue—are in between.

Some Other Features of Rainbows

The sketch displayed in figure 18.7 gives a qualitative summary of what we have covered so far. Now for a few additional remarks.

Everyone knows that the secondary rainbow is not frequently seen and even when it is visible, it is much less intense than the primary rainbow. There are two main reasons for this: (a) the light rays creating the secondary bow lose more intensity because

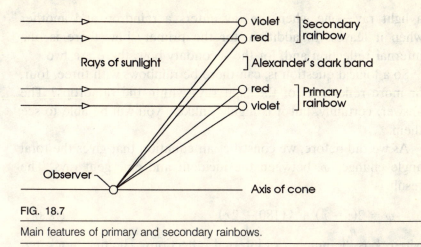

FIG. 18.7

Main features of primary and secondary rainbows.

there are two, instead of one, reflections within the raindrops and (b) as we see in a comparison of figures 18.4 and 18.6, the emergent rays of the secondary bow are much less concentrated in the vicinity of the minimum point than those of the primary bow; in other words, there is much more scattering of light in the case of the secondary bow.

Alexander's Dark Band

As we have seen, light rays of the sun are concentrated along a circular arc of about 42° radius to create the primary bow and along another circular arc of around 50° radius to produce the secondary bow. This leaves a region of about 8° in which there is much less light than elsewhere. Anyone who has observed vivid displays of rainbows has noticed this relatively dark zone between the primary and secondary bows. This zone is called Alexander's dark band, named after the Greek philosopher Alexander of Aphrodisias (c. 200 A.D.), who was evidently the first to correctly explain this feature of rainbows.

Tertiary and Quarternary Rainbows

In the preceding analysis we note that for both the primary rainbow and the secondary rainbow, there are two refractions of a light ray—one when the ray enters a raindrop and another when it leaves. In addition, for the primary bow there is one internal reflection and, for the secondary bow, there are two.

So a logical question is, can there be rainbows with three, four, or more reflections of the light rays within the raindrops? The answer: certainly, but it is highly unlikely you will be able to see them.

As we did before, we construct an equation that gives the total angle change, ϕ, between the incident and emergent rays. The result is

$$\phi = 2(i - r) + k(180 - 2r), \tag{18.8}$$

where k is the number of internal reflections. The first bracketed term on the right-hand side of this equation describes the angle changes due to the entering and departing refractions. If $k = 1$, we get equation (18.3) and if $k = 2$, we get equation (18.6).

Again utilizing (18.2) in (18.8) and then differentiating with respect to i, the following result is obtained:

$$\cos i_m = \sqrt{\frac{n^2 - 1}{k(k + 2)}}. \tag{18.9}$$

From this relationship, using $k = 1$ and $k = 2$, we regain the relations (18.5) and (18.6) we obtained earlier. Then, proceeding as before, we let $k = 3$ and $k = 4$ to provide values of i_m corresponding to the tertiary and quarternary rainbows, respectively. With this information, we compute the magnitudes of the minimum angle change ϕ_m, the corresponding cone half-angle, and so on.

An interesting result of this analysis is that both the ($k = 3$) tertiary and ($k = 4$) quarternary rainbows appear as circles around the sun! In other words, they are in the opposite direction from the primary ($k = 1$) and secondary ($k = 2$) rainbows. However, to put our minds at ease, the $k = 5$ bow coincides closely with the secondary bow and the one for $k = 6$ falls inside the primary.

Remember that these k-numbers specify the number of *reflections* of a light ray within a raindrop. The geometrical optics have been worked out for rainbows up to $k = 20$. With the help of a computer, this could be extended easily to values of k as large as we please. This might be worthwhile; interesting geometrical or numerical sequences might be generated. In any event, this topic of higher-order rainbows is well covered by Boyer (1987).

Supernumerary Bows

The contributions of many investigators, especially those of Descartes and Newton, provide a consistent and complete mechanism for the description and analysis of rainbows. This is true as far as the main features of rainbows are concerned. All we need are principles of geometrical optics including information about the variation of the refractive index with color.

Nevertheless, this relatively simple methodology cannot explain certain other features of the rainbow, including the appearance of so-called *supernumerary bows*. These rarely seen faint arcs appear near the inner edge of the primary bow and at the outer edge of the secondary.

To account for the appearance of these admittedly minor features of rainbows requires more than simple geometrical optics. It is necessary to utilize the wavelike properties of light to explain the phenomenon of supernumerary bows. Suggested references to pursue this topic are Boyer (1987) and Nussenzveig (1977). A word to the wise: fairly quickly, one can get rather deeply involved in mathematical physics on this and closely related topics.

Concluding Remarks

It is not surprising that a great deal of myth, legend, and folklore about rainbows has been generated over the centuries. An extensive coverage of such matters is given by Boyer (1987).

Along these lines, we now look at and hopefully resolve the following two topics:

The singer asserts, in the song at the start of this chapter, that bluebirds can fly over rainbows. The singer also expresses wonderment as to why he or she cannot do the same thing. Well, we simply state that it is highly unlikely, from the viewpoint of both aerodynamics and optics, that either the singer or bluebirds can accomplish this remarkable feat.

For a very long time, it has been believed that there are pots of gold at the ends of rainbows. This legend is easily dismissed. Were it true, the Internal Revenue Service, long ago, would have discovered and taxed this enormous source of income.

19

Making Mathematical Mountains

In this chapter and the one to follow, we are going to fabricate and analyze various kinds of mathematical mountains. To do this, we shall need some algebra and trigonometry and some analytic geometry and calculus. In addition, for problems with large numbers and diverse shapes of mountains, it will be necessary to utilize various topics of statistics and probability theory.

These are the basic mathematical tools used in the science of geomorphology: the study of the characteristics, origins, and changes of land forms.

Cones and Paraboloids

We begin our studies of the subject with a very simple mountain: a circular cone. What we want to do in this case, and in those to follow, is to determine the *hypsometric* (or hypsographic) curve of a particular land form, for example, a cone. Such a curve describes the mathematical relationship between the horizontal cross-sectional area of the land form and the elevation above a specified datum plane.

Our cone is shown in figure 19.1(a). The radius of the circular base of the cone is r_1 and the height is z_1. At any elevation z, the radius is r. Some analytic geometry gives the following equation

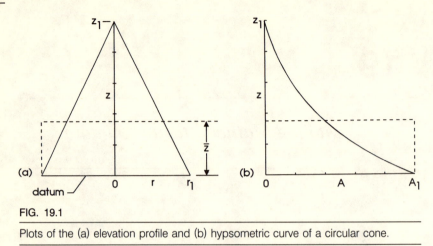

FIG. 19.1

Plots of the (a) elevation profile and (b) hypsometric curve of a circular cone.

for the edge of the cone:

$$z = z_1\left(1 - \frac{r}{r_1}\right). \tag{19.1}$$

Since the cross-sectional area is $A = \pi r^2$, the equation of the hypsometric curve is

$$z = z_1\left(1 - \sqrt{\frac{A}{A_1}}\right). \tag{19.2}$$

This relationship is plotted in figure 19.1(b).

What is the average elevation of the cone, $\bar{z}$? You may remember that the volume of a circular cone is equal to one-third the volume of the circumscribing circular cylinder, that is, $V = (1/3)A_1z_1$. So, by definition, $\bar{z} = V/A_1 = (1/3)z_1$.

Had we not remembered the one-third relationship concerning the volume of a cone, we could have computed its volume by using integral calculus. We shall do that in our next example.

However, before we leave the subject of "cones," here is an interesting little problem.

PROBLEM   If you are a pizza lover, or even if you're not, please answer the following. You cut a slice of angle α from your circular pizza and

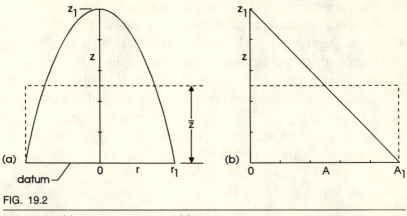

FIG. 19.2

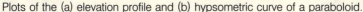

Plots of the (a) elevation profile and (b) hypsometric curve of a paraboloid.

consume it without delay at the parlor. Then you take the remaining pizza, form it into a nice cone, tape the bottom seams together, and load it up with your favorite topping at the nearby self-service counter. Fill it to a nice flat even top; no heaping on.

Question. At what angle α should you slice the pizza to provide the maximum volume of topping in your conical pizza to take home with you? Note that this problem involves some geometry and some differential calculus.

Answer. $a = 66°$.

Back to hypsometry and our next mountain. A good question: What is the mathematical shape of a land form whose hypsometric curve is a straight line? Well, from analytic geometry, we easily determine that the required equation of the hypsometric line is

$$z = z_1\left(1 - \frac{A}{A_1}\right). \tag{19.3}$$

This equation is shown in figure 19.2(b). Also, since $(A/A_1) = (r/r_1)^2$, it is easy to establish that the equation of the corresponding land form is

$$z = z_1\left(1 - r^2/r_1^2\right). \tag{19.4}$$

This equation describes an axially symmetric shape called a parabola of revolution or paraboloid. It is shown in figure 19.2(a). Beautiful Sugar Loaf peak, about 400 meters high, overlooking the harbor of Rio de Janeiro, seems to resemble a paraboloid.

Now, what is the mean elevation of our parabolic mountain? This question is easily answered by using integral calculus. We stack up an array of thin circular disks of thickness Δz and varying area A_j, to match the parabola shown in figure 19.2(a). Accordingly, the volume of the paraboloid is

$$V = \sum A_j \Delta z = \int_0^{z_1} A \, dz, \tag{19.5}$$

where, in the limit, the number of disks becomes infinitely large and the thickness of each shrinks to zero.

Solving equation (19.3) for A, substituting into (19.5), and carrying out the integration gives $V = (1/2)A_1 z_1$. In addition, we have the equation $V = \bar{z}A_1$, where, by definition, $\bar{z}$ is the average height. So we easily obtain $\bar{z} = (1/2)z_1$. As in the case of our conical mountain, $\bar{z}$ specifies the height of the melted volume.

The Great Pyramids of Egypt

On the west bank of the Nile River near Giza, not far from Cairo, are three incredible structures we shall refer to as the Great Pyramids of Egypt. They were built during the Fourth Dynasty (c. 2650 to 2500 B.C.) and are identified by the name of the kings who built them.

The main dimensions of the three pyramids are listed in table 19.1. These are the estimated original dimensions; not surprisingly, 4500 years has caused considerable erosion. The pyramids are listed in the table in chronological order of construction.

Now what we are going to do, perhaps for the first time in history, is determine the hypsometric curve of the Great Pyramids. This is an interesting problem and one that forms the basis for more complicated ones we consider later.

It is assumed that the three pyramids have the same datum plane, $z = 0$. For pyramid 1, Cheops, we write the profile equa-

TABLE 19.1

The Great Pyramids of Egypt

No.	Name	Height m	Base length m	Base area m²	Hypotenuse m
1	Cheops	147	236	55,696	189
2	Chephren	144	216	46,656	180
3	Mycerinus	66	109	11,881	86

Source: Edwards (1993).

tion, $z = z_1(1 - r/r_1)$, where z_1 is the height and r_1 is the base length. Since the horizontal sections of the pyramids are squares, we have $A/A_1 = (r/r_1)^2$ and so $A = A_1(1 - z/z_1)^2$, where A_1 is the base area. Therefore, for the three pyramids the hypsometric equation is

$$A = A_1\left(1 - \frac{z}{z_1}\right)^2 + A_2\left(1 - \frac{z}{z_2}\right)^2 + A_3\left(1 - \frac{z}{z_3}\right)^2. \quad (19.6)$$

For the single-cone case, given by equation (19.2), we expressed our hypsometric equation in the explicit form, $z = f(A)$, instead of the implicit form, $A = f(z)$. The explicit form is usually preferable if we want to compute slopes or calculate volumes. If we have two cones or two pyramids, it is still fairly easy to obtain the explicit form although we get some messy quadratic equations. For three or more peaks, the tedious algebra is not worth the trouble.

Accordingly, we regard equation (19.6) as the final result. Putting the numerical values listed in table 19.1 into this equation and carrying out the simple calculations produces the hypsometric curve shown in figure 19.3(b). The half-profile of the corresponding equivalent single structure is displayed in figure 19.3(a).

Utilizing the fact that the volume of a pyramid is one-third the volume of the containing rectangular prism, the average height of the three pyramids is $\bar{z} = 45.8$ m. The total volume $V_T =$

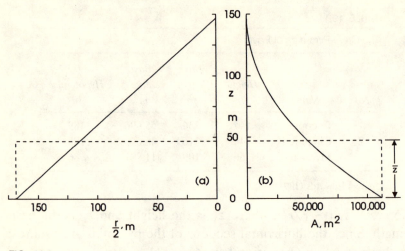

FIG. 19.3

The Great Pyramids of Egypt. (a) Equivalent single structure (half base only); total base length $r_e = 338$ m, total base area $A_T = 114{,}233$ m^2, height $z = 147$ m, average height $\bar{z} = 45.8$ m. The structure is slightly concave upward. (b) Hypsometric curve.

5.23 million m^3. This is about twice the volume of Hoover Dam on the Colorado River in Arizona–Nevada.

The ancient Egyptians had an incredible ability in civil engineering. Cheops is nearly as high as the United Nations building (thirty-nine stories) in New York and its base covers the area of fifteen football fields—a fantastic engineering accomplishment, even by present-day standards. And it was built when power meant humans and animals!

The Egyptians also had a surprisingly large knowledge of mathematics and astronomy. They had computed the circumference of the earth to nearly modern-day accuracy and they knew that the value of π was around 3.1. In their structural and irrigation engineering projects, it is possible that they were aware of the right-triangle relationship: $3^2 + 4^2 = 5^2$.

Here is another example. In various other chapters we have met the interesting quantity called the divine proportion or so-called golden number, $\phi = (1/2)(1 + \sqrt{5}) = 1.61803$. The ancient Egyptians may have known about this number also. It has been suggested by Ghyka (1978) that the geometries of the Great Pyramids feature the ratios

$$\frac{\text{half base length}}{1} = \frac{\text{height}}{\sqrt{\phi}} = \frac{\text{hypotenuse}}{\phi}.$$

Computing the height and hypotenuse of each of the pyramids using these relationships and the base lengths shown in table 19.1, yields results surprisingly close to the actual values.

Over the centuries, just about everything involving engineering, science, and technology—from Egyptian pyramids and medieval cathedrals to supersonic aircraft and maglev trains—has utilized mathematics to one extent or another. The history of mathematics is fascinating and there are many excellent books devoted to the subject. Two that are especially readable are those of Boyer (1991) and Resnikoff and Wells (1984). In addition, there is an interesting four-volume set of books edited by Newman (1956) entitled *The World of Mathematics* and an equally interesting three-volume set by Campbell and Higgins (1984) called *Mathematics*: *People*, *Problems*, *Results*. All of these references are recommended to those who would like to learn more of the history of mathematics.

Looking Ahead: Mountain Ranges and Molehills

This is an appropriate place to pause in our study of geomorphology and hypsometric curves. We continue with these topics in the next chapter, in which we build a mathematical mountain range. For this, as the title of the chapter indicates, there will be need for a certain number of molehills. Once we obtain the mathematical description of such a mountain range, we can calculate its hypsometric curve.

20

How to Make Mountains out of Molehills

In the preceding chapter, in an introduction to the subject of geomorphology, the so-called hypsometric curve was defined and several examples were computed. So we ask the question: Why do geomorphologists study things like hypsometric curves and related topics? The following paragraphs may provide some answers.

Our Always-Changing Planet

Geomorphology deals with the characteristics, origins, and changes of land forms. The incredibly powerful forces of nature—aided and abetted by mankind these days—continually alter the topography and geography of our world. Rainfalls and snowfalls, floods, droughts, winds, tornadoes, hurricanes, typhoons, landslides, avalanches, earthquakes, volcano eruptions, forest fires, ocean waves, and tides—not to mention glacier movements and periodic ice ages—never stop changing the physical features of our earth.

These forces determine the extents of soil erosion, surface runoff and drainage, mud slides and debris flows, river meandering and delta formation, reservoir siltation and waterway clogging.

All these things are better understood and dealt with when we study land mass shapes and their never-ending changes. Hypsometric curves—the quantitative relationships between land area and elevations above a datum—provide valuable information for projects dealing with agriculture and irrigation, city and regional planning, highway and airport location, river and coastal engineering, and so on.

Mathematical modeling is utilized by geomorphologists to examine these numerous land-changing phenomena. An interesting model for the study of the hydrology of drainage basins was devised by Chorley and Morley (1959). They fabricated a mathematical basin by selecting, as their mountain, a cylinder of lemniscate cross-section whose equation is

$$(x^2 + y^2)^2 = a^2(x^2 - y^2),$$ (20.1)

which looks something like a narrow leaf or a pointed circle. They intersected this solid with a circular cone which has the shape equation

$$x^2 + y^2 = c^2 z^2.$$ (20.2)

In these expressions, (x, y, z) are the rectangular coordinates and a and c are constants. It turns out that the region contained by the two spatial surfaces, equations (20.1) and (20.2), looks much like an actual drainage basin. These researchers needed to calculate the hypsometric curve in order to study the hydrology and drainage characteristics of the basin.

Here is another example. In his analysis of the hypsometry of the continents, Cogley (1985) employed computer models to analyze enormous amounts of data. In chapter 19, we looked at topography involving one cone, one paraboloid, and three pyramids; shortly we will construct and analyze a mountain range composed of 385 well-defined peaks. However, Cogley was looking at the topography of the entire world. Consequently, in order to handle extremely large amounts of information, he had to use many methods and techniques of statistics and probability. A comprehensive, though rather advanced, reference on the use of

these methodologies in geomorphology is the book edited by
Chorley (1972).

A Mathematical Mountain Range

For the mountain range we want to construct, we will employ a
more general form of the equation we used in our analysis of the
three Great Pyramids of Egypt. That is,

$$A = \sum_{j=1}^{m} A_j \left(1 - \frac{z}{z_j}\right)^2, \tag{20.3}$$

where, for the three pyramids, $m = 3$. It is supposed that all our
mountains—or hills, mounds, heaps, piles, and molehills—are
pyramids with square cross-sections and the same base area. The
total base area is A_T. There are n_j peaks in a certain category of
designated height z_j, and there are m height categories.

So, letting $n_j = A_j/A_T$, equation (20.3) becomes

$$\frac{A}{A_T} = \sum_{j=1}^{m} n_j \left(1 - \frac{z}{z_j}\right)^2 \bigg/ \sum_{j=1}^{m} n_j. \tag{20.4}$$

This equation is not as formidable as it appears, as we shall see in
a moment. We select the mountain range design recipe (MRDR)
shown in table 20.1. In our design recipe, the *number* of peaks in
each category is increased in a squared sequence. The *heights* of
the peaks in successive categories are decreased in a linear
sequence. Altogether there are 385 peaks including, as shown in
the table, one big mountain, 9 hills, ..., 64 piles, and 100
molehills.

Consequently, our hypsometric equation is

$$\frac{A}{A_T} = \frac{1}{385} \left[1\left(1 - \frac{z}{z_1}\right)^2 + 4\left(1 - \frac{z}{0.9z_1}\right)^2 \right.$$

$$\left. + \cdots + 100\left(1 - \frac{z}{0.1z_1}\right)^2 \right]. \tag{20.5}$$

TABLE 20.1

Mountain range design recipe

j	n_j	z_j / z_l	Object
1	1	1.0	mountains
2	4	0.9	
3	9	0.8	hills
4	16	0.7	
5	25	0.6	mounds
6	36	0.5	heaps
7	49	0.4	
8	64	0.3	piles
9	81	0.2	
10	100	0.1	molehills

The average elevation, $\bar{z}/z_1$, is obtained from the expression

$$\frac{\bar{z}}{z_1} = \sum_{j=1}^{m} n_j \left(\frac{z_j}{z_1}\right) \Big/ 3 \sum_{j=1}^{m} n_j. \tag{20.6}$$

For our particular mountain range, $\bar{z}/z_1 = 0.1048$.

Results calculated from equation (20.5) are shown in figure 20.1. For comparison, the hypsometric curve of the earth's land surface is also shown. Using $z_1 = 8{,}847$ m (Mount Everest) and $\bar{z} = 840$ m, we obtain $\bar{z}/z_1 = 0.0949$. It is noted that the hypsometric curve of the simple mountain range we just now manufactured is not greatly different from that of the earth. Data provided by Cogley (1985) were utilized to construct the latter curve.

As we did in the case of the Great Pyramids of Egypt, we construct a single mountain equivalent to our 385-peak mountain range. For a change, we assume that the equivalent mountain has a circular cross-section. Likewise, we construct an equivalent circular mountain corresponding to the earth's total land mass. The profiles of these two equivalent peaks are shown in figure 20.2.

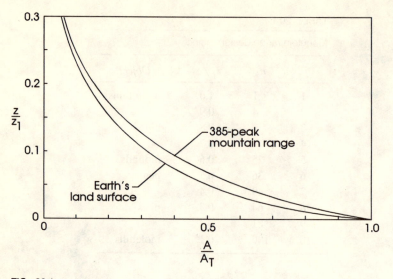

FIG. 20.1

Hypsometric curves for our 385 peak mountain range and for the earth's total land surface. (From Cogley 1985.)

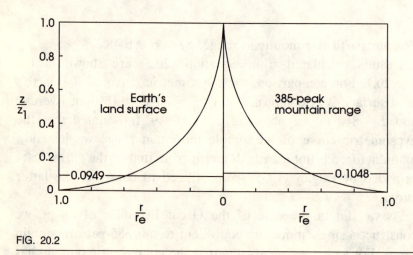

FIG. 20.2

Elevation profiles of equivalent single axially symmetric peaks. *Right:* our 385-peak mountain range. *Left:* the earth's total land mass with $z_1 = 8,847$ m (Mount Everest) and $A_T = 1.49 \times 10^8$ km^2, corresponding to an equivalent radius $r_e = 6,887$ km on a plane surface. Average elevations for our 385-peak mountain range and the earth's land surface are also shown.

PROBLEM 1.   Design your own mountain range using whatever design recipe (MRDR) you like. For the shape of the components of your mountain range you might want to use rectangular boxes, circular cylinders, or cones. Perhaps you prefer pyramids, paraboloids, or something else. For the height categories and the number of peaks within categories, you might want to use (a) an arithmetic progression, (b) a geometric progression, (c) binomial numbers, (d) factorial numbers, (e) Fibonacci numbers, (f) hailstone numbers, or (g) random numbers.

PROBLEM 2.   Compute and plot the hypsometric curve and the elevation profile of the equivalent single-peak axially symmetric mountain. If you like to do things with computer graphics, perspective displays of your mountain range would be very interesting.

A Brachistochrone Mountain

Quite a long time ago, in 1696 to be precise, the famous Swiss mathematician Johann Bernoulli (1667–1748) posed the following problem:

> Due to the force of gravity, a particle, initially at rest, slides without friction along a curve in a vertical plane from point P to point Q. What should be the shape of the curve to assure minimum transit time between the two points?

As was the custom in those days, mathematicians throughout Europe competed for the honor to provide answers to such problems. One solution to this problem was given by Jakob Bernoulli (1654–1705), the older brother of Johann, and another was submitted by the well-known German mathematician and coinventor of calculus, Gottfried Leibniz (1646–1716). The famous English mathematician and other coinventor of calculus, Isaac Newton (1643–1727), also solved the problem. Indeed it is said that the day after he received the problem, Newton sent the solution to a friend.

The answer to this intriguing problem is that a curve called the *brachistochrone* is the curve that provides quickest descent. Indeed, the name itself is derived from the Greek words *brachystos* (shortest) and *chronos* (time). As Newton pointed out, this curve is the same thing as the *cycloid*, which is the path described by a point on the periphery of a circle of radius a as it moves along a straight line.

In parametric form, the cycloid is defined by the equations

$$x = a(\theta - \sin \theta); \quad y = a(1 - \cos \theta), \tag{20.7}$$

where θ is the parameter, whose value ranges from $\theta = 0$ to $\theta = \pi$.

To obtain the answer expressed by equations (20.7), we utilize the branch of mathematics called the *calculus of variations*. This provides a method that mathematicians employ to determine the forms of functions providing maximum or minimum values of things. In the brachistochrone problem, minimum time is the quantity we are examining. In other problems, minimum length, minimum area, minimum energy, or something else might be the variable of interest. A suggested reference on the subject is the fairly easy to read book by Forray (1986).

Since the main subject of this chapter is mathematical mountains, it is appropriate that we construct a brachistochrone mountain. To do so, we need to modify (20.7) to the form

$$\frac{r}{a} = \theta - \sin \theta; \quad \frac{z}{a} = 1 + \cos \theta, \tag{20.8}$$

where (r, z) are, respectively, the radial and vertical coordinates of the profile of our axially symmetric mountain. The coordinates of the top of the mountain are $(0, z_0)$ and the coordinates of the base are $(r_*, 0)$. It is easy to show that $z_0 = 2a$ and $r_* = \pi a$, where, again, a is the radius of the circle that generates the cycloid.

If we like, we can eliminate the parameter θ by using the second of equations (20.8) to obtain an expression for θ and substituting the answer into the first equation. The result is

$$\frac{r}{a} = \arccos\left(\frac{z}{a} - 1\right) - \sqrt{2\left(\frac{z}{a}\right) - \left(\frac{z}{a}\right)^2}. \tag{20.9}$$

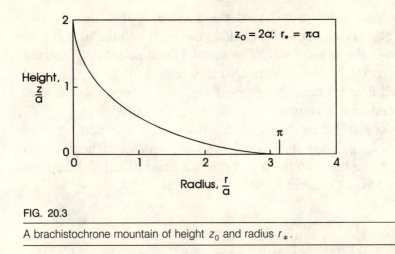

FIG. 20.3

A brachistochrone mountain of height z_0 and radius r_*.

This is an alternative form of the brachistochrone curve. A plot of (20.8)—or the equivalent (20.9)—is displayed in figure 20.3.

Can you show, using either (20.8) or (20.9), that the *slope* of this curve, dz/dr, is infinite at the top of the mountain $(0, z_0)$ and zero at the base $(r_*, 0)$? Accordingly, we would say that a *cusp* exists in the curve at the top of the mountain.

Now we can calculate a few things about our brachistochrone mountain.

Arc Length

Starting with the relationship $ds^2 = dr^2 + dz^2$ and utilizing the parametric relationships of equations (20.8), it is not difficult to establish that the length of the curve from the top of the mountain to the base is $S = 2z_0$.

Time of Descent

The time of descent of an object sliding from the top to the base is given by the equation

$$T = \pi \sqrt{\frac{z_0}{2g}} \, . \tag{20.10}$$

An interesting property of the cycloid is that the time of descent from *any* point (r, z) along the profile to the base $(r_*, 0)$ is always this same value. The noted Dutch scientist Christiaan Huygens (1629–1695) discovered this feature of the cycloid and utilized it in his tautochrone clock. In such a clock the period of the pendulum is constant regardless of the amplitude.

An object falling freely in a vacuum has the descent time $T = \sqrt{2z_0/g}$. Show that an object sliding without friction along a straight line between the top of the mountain $(0, z_0)$ and the base $(r_*, 0)$ has the descent time

$$T = \sqrt{1 + \left(\frac{r_*}{z_0}\right)^2} \sqrt{\frac{2z_0}{g}}. \tag{20.11}$$

At this point it may be helpful to introduce some numbers. Suppose you are ready to go on your frictionless skis at the top of our snow-covered brachistochrone mountain of height $z_0 = 100$ m, and radius $r_* = (\pi/2)z_0 = 157$ m. As we established above, the arc length of your trip from the top to the base is $S = 2z_0 = 200$ m.

From (20.10), the time of your descent is $T = \pi\sqrt{z_0/2g} = \pi\sqrt{100/2(9.82)} = 7.09$ s. Accordingly, your average velocity is $\overline{U} = S/T = 200/7.09 = 28.2$ m/s. Compare this to your velocity at the instant you reach the base, $U_* = \sqrt{2gz_0} = 44.3$ m/s.

If somehow you could have gone in a straight line from the top to the base, the distance would have been $S = \sqrt{z_0^2 + r^2_*} = 186$ m. From equation (20.11), your time of descent would have been $T = 8.40$ s with an average velocity $\overline{U} = 186/8.40 = 22.1$ m/s. So even though the straight-line path is shorter than the brachistochrone path (186 meters vs. 200 meters), the straight route takes longer (8.40 seconds vs. 7.09 seconds).

You might want to examine the times of descent T, arc lengths S, and average velocities $\overline{U}$, of some other descent curves. Parabolic, circular, and sinusoidal paths would be interesting cases to consider.

FIG. 20.4

The Mayon Volcano in the Philippines. (Photograph provided by Embassy of the Philippines, Washington.)

Numerous presentations about the history of the brachistochrone—and many other stories concerning the development of mathematics—are given by Struik (1967) and by Dunham (1990).

The Mayon Volcano in the Philippines

Without doubt the most perfectly symmetrical mountain in the world is beautiful Mayon Volcano in the province of Albay in the southern part of Luzon in the Philippines. This peak, shown in figure 20.4, has an elevation of 2,460 meters and covers an area of over 300 square kilometers. It has a long history of eruption. The first recorded event was in 1616. There have been over fifty eruptions since then, including one in 1814 that completely

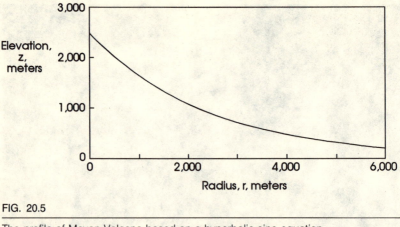

FIG. 20.5

The profile of Mayon Volcano based on a hyperbolic sine equation.

buried the town of Cagsawa with substantial loss of life. There was intense activity of the volcano in 1993.

This mountain is so symmetrical that, over the years, many mathematically inclined people have attempted to describe its shape in terms of mathematical equations. Among the first to do so was Becker (1905), who decided the peak looked like a hyperbolic sine curve. Specifically, he employed the equation

$$\frac{r}{c} = \sinh\left(\frac{z_0 - z}{c}\right),$$ (20.12)

where the scaling constant c has the numerical value $c = 800$ m. Others have matched the volcano's lovely profile to exponential curves, cosine curves, and logarithmic curves.

To conclude our chapter on mathematical mountains, we take a closer look at the hyperbolic sine relationship expressed by equation (20.12). The elevation of Mayon Volcano is $z_0 = 2,460$ m. Substituting this quantity and the value $c = 800$ m into (20.12), we compute the profile $z = f(r)$. The result, shown in figure 20.5, is displayed to the same scale as the photograph in figure 20.4.

PROBLEM 1. It is appropriate that we close this chapter with the assignment that you compute the hypsometric curve for (a) a brachistochrone mountain and (b) a hyperbolic sine mountain.

PROBLEM 2. Finally, here is a little problem designed to help prepare us for the next chapter. A hunter leaves his home one morning and goes straight south for one mile. Then he turns 90° and goes straight east for one mile. At that point he sees a bear and takes a shot at it. Wisely, he decides to go directly home. When he gets there, he determines that his entire trip was exactly three miles.

What color was the bear?

21

Moving Continents from Here to There

What Color Was the Bear?

It turns out the bear was white, because it was a polar bear, because the hunter lived right at the North Pole, because only at the North Pole could one carry out the geometrical journey described in the question. The crux of the matter, as shown in figure 21.1, is that we are dealing with a *spherical* triangle, not a *plane* triangle. You might want to examine this geometry on your globe of the world or your basketball.

When dealing with problems involving distances on the earth's surface, our computations are simplified if we use *nautical miles*, instead of *statute miles* or *kilometers*. The nautical mile is explained in the following paragraph.

We assume that the earth is a perfect sphere. Lines of constant longitude passing through the North and South Poles—the great circles called meridians—can be divided into 360 degrees; each degree consists of 60 minutes of latitude change. By definition, the arc length along a meridian corresponding to one minute of latitude difference is one nautical mile. The maritime unit of velocity, the knot, is equivalent to one nautical mile per hour.

Clearly, the circumference of the world is $C = (360)(60) = 21,600$ nautical miles. It can be established that 1.0 nautical mile = 1.1510 statute miles = 1.8520 kilometers. On this basis,

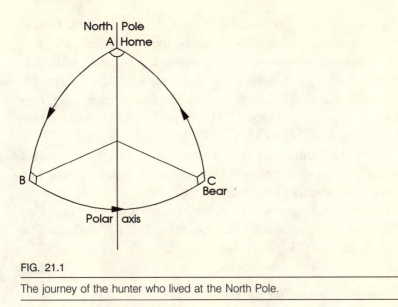

FIG. 21.1

The journey of the hunter who lived at the North Pole.

the radius of the earth is $R = C/2\pi = 3{,}437.7$ nautical miles = $3{,}956.2$ statute miles = $6{,}366.7$ kilometers.

How far "around the world" did the hunter go on the west-to-east (B to C) leg of his journey just before he spotted the bear? It is helpful to refer to figure 21.2 to answer this question.

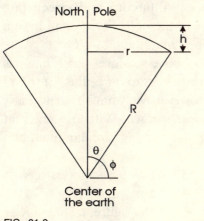

FIG. 21.2

Definition sketch for spherical geometry involving the North Pole.

It is seen in the figure that $\sin\theta = r/R$. On his west-to-east segment, the hunter was walking along the line of latitude $\phi = 89°59'$N. Accordingly, $\theta = 1/60°$. Letting $C_* = 2\pi r$ be the circumference of this latitude circle, we obtain $C_* = C\sin\theta = 21,600\sin(1/60) = 6.2832$ nautical miles. Since the hunter covered a west-to-east distance of 1.0 nautical miles, it follows that he went $(1.0/6.2832)(360°) = 57.296°$ around the world.

Suppose the hunter, as he left his home and went straight south on leg A to B, was in fact headed directly toward San Francisco, California (longitude $\lambda = 122°$W). Then just after he took a shot at the bear and went straight north toward home on leg C to A, he was headed directly away from Moncton, New Brunswick ($\lambda = 65°$W). It's a small world.

Wherein We Construct an Enormous Mountain at the North Pole

The main reason we came to the North Pole was to supervise the moving of the earth's entire land mass to the region of the North Pole. The authorities inform us that it is necessary to arrange the land mass so that it is identical to the axially symmetric peak shown in figure 20.2. We put Mount Everest (elevation 8,847 m) at the center, on the polar axis.

Bear in mind that we also move Antarctica and Greenland, with all their ice caps and glaciers, to the pole. So we shall have a truly beautiful ice- and snow-capped mountain.

The total area of the world's above-sea-level land mass is $A_c = 1.49 \times 10^8$ km^2. This corresponds to a radius of 6,887 kilometers on a *plane* surface. However, we want the land mass to fit symmetrically onto a *spherical* surface. With its center at the North Pole, to what latitude will our big circular mountain extend southward?

With reference to figure 21.2, the area of a spherical cap is $A_c = 2\pi Rh$. Using $A_c = 1.49 \times 10^8$ km^2 and $R = 6.370 \times 10^3$ km, we obtain $h = 3,723$ km. From the figure, $\cos\theta = (R - h)/R$. This gives $\theta = 65.45°$ and so the latitude at the edge of our circular land mass is $\phi = 90° - \theta = 24°33'$N. The tropic of Cancer is located at $\phi = 23°27'$N; consequently, the land mass extends almost to the tropics. Some cities at or near the edge of our circular island are Key West, Florida (24°33'), Honolulu, Hawaii (21°18'), Brownsville, Texas (25°54'), Riyadh, Saudi Arabia (24°40'), Karachi, Pakistan (24°52'), and Taipei, Taiwan (25°03').

What is the length of the shoreline of the circular island? If the land mass were on a plane surface the circumference would be $C_* = 2\pi(6,887) = 43,272$ km. However, on a spherical surface, $C_* = 2\pi r = 2\pi R \sin \theta = 36,406$ km. Clearly, this is the circumference of a circle located at latitude $\phi = 24°33'$.

Obviously, the remaining and larger area of the world is covered by water. Thank goodness the two areas—lands and oceans—are not equal. Otherwise it would be necessary to resolve the following troublesome dilemma:

> If the entire northern hemisphere were land and if the entire southern hemisphere were water, would we have an island in the center of a large body of water or a lake in the center of a large mass of land?

Freezing Water and Melting Ice

During the past several million years, the climate of the earth has alternated between cold periods resulting in ice formation and glacial advance and warm periods causing ice melting and glacial retreat.

At the height of the most recent "ice age"—about twenty thousand years ago—vast sheets of ice covered much of Alaska, all of Canada, Greenland, Iceland, and the British Isles, and a good deal of the northern regions of Europe and Asia. In the United States, enormous ice sheets and glaciers extended over much of the North Central, Middle Atlantic, and New England states.

It has been estimated by Denton and Hughes (1981) that the total volume of ice during the ice age was about 90 million cubic kilometers—three times more than we presently have, mostly in

Antarctica and Greenland. That enormous volume would provide an ice blanket 4,800 meters thick, about three miles, over Alaska, Canada, and the forty-eight states of the United States.

Of course, all the water forming these vast ice sheets came from the oceans. During the ice age, volumes of frozen ocean water were so gigantic that, according to Gross (1985) and Byalko (1987), the level of the ocean was approximately 120 to 150 meters below the present level. Most of what we now call the relatively shallow "continental shelf" was dry land.

There is a fascinating anthropological-geographical feature related to the great ice age of twenty thousand years ago. The Bering Strait is a relatively narrow (88 kilometers) passage connecting the Pacific and Arctic Oceans, separating Siberia from Alaska. Even now the maximum depth of the Strait is only 50 meters. During the ice age there was certainly a land bridge, perhaps 100 meters above sea level, connecting Asia and North America.

Anthropologists believe that tribes of people from Northern Asia crossed this bridge, over time periods of hundreds to thousands of years, and slowly migrated eastward and southward throughout North, Central, and South America. In this regard, in their very interesting book dealing with the Mayan civilization of Mexico, Schele and Freidel (1990) indicate that "... archaeological history begins with evidence of the first people moving into the Yucatán Peninsula about eleven thousand years ago." The "Bering bridge" hypothesis is certainly plausible and an interesting thing to think about.

Back to the ice-covered circular land mass we built at the North Pole. It is estimated that the present volume of ice in the earth's glaciers and ice caps is about 30×10^{15} m^3. Although this amount represents only about 2.1% of all the water in the world, these ice caps and glaciers contain nearly 80% of the earth's *fresh* water.

According to Oerlemans and van der Veen (1984), if all this ice were to melt, the level of the oceans would rise by approximately

72 meters. These researchers estimate that the current rate of melting is sufficient to cause an ocean level rise of about 0.5 centimeters per year or about 50 centimeters per century. At this rate, it would take about 15,000 years for all the ice to melt. It is highly probable that the melting rate will increase if greenhouse warming becomes a serious problem.

In addition to the melting of ice, global warming would have another significant effect on raising ocean levels: thermal expansion of sea water. For example, if the temperature of all ocean water were suddenly increased from 15°C to 20°C, the level would rise by about 3.5 meters because of thermal expansion of the water.

There is certainly not complete agreement among scientists regarding the long-term rate of glacier and ice cap melting, so we set that matter aside. However, there is no question concerning the enormous annual production of ice. It has been estimated by Oerlemans and van der Veen that about 3.0×10^{12} m^3 of ice are produced annually, mostly in Antarctica (80%), a great deal in Greenland (15%), and the remainder in the Canadian and Russian Arctic islands.

Approximately 10% of this annual ice production is lost to the oceans by outright melting during the warmer seasons. However, nearly 90% of the total—and virtually all of Antarctica's ice—falls into the ocean by calving: the breaking off of large detached pieces. This is the process by which enormous icebergs are created.

The annual production and subsequent thawing of 3.0 million cubic meters of ice each year involves a great deal of frozen and then melted water. The corresponding melting rate is fifteen times larger than the flow over Niagara Falls and five times more than the discharge of the Mississippi River. It would fill Lake Erie in two months. It is double the total consumptive use of water in the entire world.

It is important to remember that all this water from glaciers and icebergs, before it mixes with ocean water, is fresh water. What a dreadful shame that we are unable to use it.

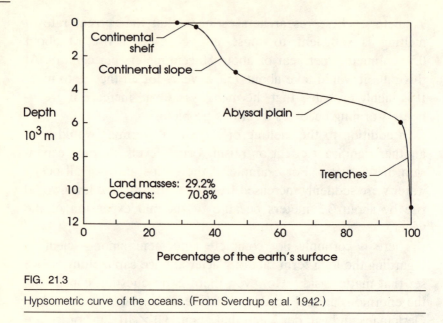

FIG. 21.3

Hypsometric curve of the oceans. (From Sverdrup et al. 1942.)

What Happens If We Dump All the Land into the Ocean?

The hypsometric (or bathometric) curve of the oceans is shown in figure 21.3. As indicated previously, the world's land mass covers an area $A_c = 1.49 \times 10^{14}$ m² (29.2% of the total) and the world's oceans extend over an area $A_o = 3.61 \times 10^{14}$ m² (70.8%).

As shown in the figure, the hypsometric curve features a continental shelf to a depth of about 135 meters. This area was above sea level during the ice age. The shelf is followed by a zone called the continental slope to a depth of around 3,000 meters. From this depth to approximately 6,000 meters is a vast abyssal plain or deep sea platform.

Finally, the hypsometric curve displays a relatively small region containing trenches with very large depths. The greatest known depth is 11,035 meters in the Marianas Trench south of Guam in the Pacific Ocean.

The average depth of the oceans is $\bar{z}_o = 3,740$ m. Consequently, the volume of the oceans is $V_o = 1,350 \times 10^{15}$ m³. The average elevation of the earth's land mass is $\bar{z}_c = 840$ m and so the volume is $V_c = 125 \times 10^{15}$ m³.

Recall that we had moved the entire above-sea-level land mass to the North Pole to construct an axially symmetric mountain conforming exactly to the hypsometric curve of the earth's continents. The volume of the land mass was $V_c = 125 \times 10^{15}$ m³. Along with the land, we moved all of the earth's glaciers, ice caps, lakes, and rivers with a total volume $V_w = 37.5 \times 10^{15}$ m³. So the entire amount moved was $V = 162.5 \times 10^{15}$ m³.

After our mountain construction project is completed, the decision is made to dump the entire mass—dirt, rock, sand, ice, everything—into the ocean. What happens?

The total area of the world is $A = 5.10 \times 10^{14}$ m². So the enormous land mass that we dump into the ocean, along with all the ice and fresh water, raises the ocean level by an amount $h = V/A = (125 + 37.5)(10^{15})/(5.10)(10^{14}) = 318$ m.

However, this dumping operation still leaves an area, formerly covered by land, where the depth is only 318 meters, and a much larger area where the depth is much more than this. So, in an effort to be neat and orderly, we launch a massive project of underwater bulldozing to level off the entire bottom of the oceans.

The outcome is that our earth is now a perfectly smooth sphere covered entirely by an ocean with a constant depth of $h = V/A = (1,350 + 37.5)(10^{15})/(5.10)(10^{14}) = 2,720$ m. All is serene.

22

Cartography: How to Flatten Spheres

It is said that Columbus must surely have been an economist, a stockbroker, or something like that to have successfully raised the funds for the journey of his three small ships across the Atlantic back in 1492. But even though Columbus did receive fairly strong financial support from the Spanish crown, he definitely was not an economist. In fact, he was a very competent seaman and, more importantly, he possessed considerable knowledge of and experience in the principles of navigation and geography.

The fact that Columbus mistook the vast land mass we now call the Americas for south or east Asia was not entirely his fault. By the late fifteenth century, when he made his journeys, there were quite accurate maps of most of Europe and the Middle East and fairly good maps of much of Africa and Asia. However, the New World simply did not exist!

Indeed, much of the geography of the Old World had been known since the time of the great astronomer Claudius Ptolemy of Alexandria (87–150). Even before that, the Babylonians, Persians, and Egyptians had produced maps of the then-known world and maps of the skies. By the fifth century, the Greeks had laid the foundations of what we now call *cartography*: the science and art of maps.

Over the centuries, a great many people made contributions to the development of maps and globes of the world and the heavens. Two names especially stand out in the long history of cartography. One is Ptolemy and the other is Gerhard Kremer (1512–1594); we know him better by his latinized name, Gerardus Mercator. Though born in Holland, he spent nearly all of his long life in Germany. In 1569 his famous *Great World Map* was published. Mercator's map, or more precisely, the Mercator projection of the world, is familiar to all of us; we shall return to it shortly.

There are numerous books devoted to the history of maps and cartography; those of Bagrow (1985) and Brown (1949) are recommended. *Maps of the Heavens* is the title of a beautiful book by Snyder (1984) dealing with cartography of the skies.

How Map Projections Are Classified

Suppose we have a globe or some other spherical object on which we have a map of the world, or indeed any kind of configuration or pattern of lines. Suppose also that we want to transform this three-dimensional sphere to a two-dimensional plane in order to produce a map of the world or pattern of lines on a nice sheet of flat paper.

Basically there are two methods to accomplish this. The first method is to place the sphere—the globe—under a powerful hydraulic press, like those used in steel mills, and flatten the heck out of it. We discard this procedure because it is not scientific. The second method is to utilize various techniques to transform the sphere mathematically onto a plane.

It turns out that it is impossible to transform a sphere to a plane without some kind of distortion in the map you are making. If you want to retain some features of the map in the transformation from sphere to plane (e.g., angles or distances), then you must sacrifice other features (e.g., areas or directions).

There are three types of surfaces we can use to transform or project a three-dimensional sphere onto a two-dimensional plane.

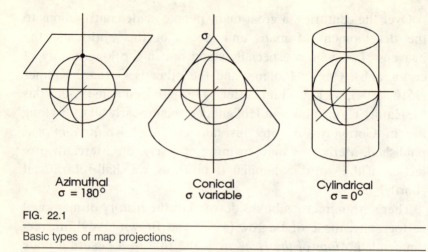

| Azimuthal
σ = 180° | Conical
σ variable | Cylindrical
σ = 0° |

FIG. 22.1

Basic types of map projections.

These are illustrated in figure 22.1. The first is a plane surface itself; this is called the *azimuthal* projection. The second is the *conical* projection. In this case, after the sphere's pattern is projected onto the cone, a cut is made along the side of the cone and the surface is spread out to provide a flat map. The third is the *cylindrical* projection. Again, after the globe's pattern is projected, the cylinder is cut and laid flat.

As shown in the figure, these three basic geometries, the plane, cone, and cylinder, are *tangent* to the sphere at a certain point or along a certain curve. However, each of the three surfaces could actually intersect the sphere at a specified place; these are called *secant* projections. Although there are many applications for secant maps, we shall not consider them here.

From the preceding, we note that the first way to classify map projections is according to the type of plane surface onto which the sphere is projected: azimuthal, conical, or cylindrical.

A second way to classify projections is according to the projection source. For example, suppose we want to construct an azimuthal projection. That is, as shown in figure 22.2, we want to project point P of the sphere onto the plane. We could select the center of the sphere, point A, as the projection source and obtain point P_1 on the plane. This is called a *gnomonic* projection. Alternatively, the projection source could be at point B,

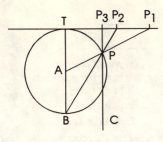

FIG. 22.2

Types of projection sources. (a) APP_1: gnomonic. (b) BPP_2: stereographic. (c) CPP_3: orthographic.

located directly opposite the point of tangency T. This is called the *stereographic* projection; it produces point P_2 on the plane. Finally, we could move the projection source all the way to infinity at point C to generate point P_3 on the map. This is the *orthographic* projection.

There is a third way by which map projections can be classified. This classification specifies which *properties* we want to preserve as we transform from a sphere to a plane. There are four such properties:

Equal angles. This property, usually called the conformality property, assures that any angle on the sphere is transformed without change onto the plane. For example, the meridians (lines of constant longitude) and the parallels (lines of constant latitude), which are perpendicular on the sphere, are also perpendicular on the plane.

Equal areas. This property stipulates that every small region on the sphere has the same area after it is transformed onto the plane.

Equal distances. This property requires that distances from the center of a map projection be the same on the sphere and the plane.

True directions. This property means that directions from the center of a map projection must be identical on the surface and the plane.

As indicated above, it is not possible to preserve all these properties in a particular transformation. There must be trade-offs. For example, the flag of the United Nations displays an

TABLE 22.1

Classifications of map projections

Projection plane	Projection source	Preserved property
Azimuthal	Gnomonic	Equal angles
Conical	Stereographic	Equal areas
Cylindrical	Orthographic	Equal distances
		True directions

azimuthal stereographic conformal (equal-angle) projection of the world with the point of tangency at the North Pole. The other properties are not preserved on this projection. Table 22.1 gives a summary of the various ways by which map projections are classified.

Over the past two thousand years or so, many hundreds of different kinds of map projections have been devised. Of this very large number, perhaps fifty kinds find some type of present-day application and maybe a dozen or so can be described as extremely useful. Not surprisingly, mankind's relatively recent ventures into space exploration have produced renewed interest and great progress in mapping not only the earth but also the moon and nearby planets.

We are going to take a close look at only two of the many different kinds of projections. If you want to study cartography and map projection in detail, numerous books are available; some of the best are those of Raisz (1962), Snyder (1987), Snyder and Voxland (1989), and Snyder (1993).

In your studies, it will be very helpful if you know something about spherical trigonometry and solid analytic geometry. If you would like to become knowledgeable in cartography, you should study calculus and differential equations and also an interesting branch of mathematics called differential geometry.

The two projections we are going to examine in some detail are the Mercator projection and the Lambert azimuthal equal-area projection. We start with the first of these.

The Mercator Projection and Loxodromes

This projection was devised by the great Flemish cartographer Gerardus Mercator (1512–1594) and first appeared in 1569. It is developed onto a cylinder in such a way that angles are preserved (i.e., it is *conformal*). Without doubt, this is the most famous of all map projections. It has the great advantage that paths on a sphere that hold constant angle to the meridians (lines of constant longitude) transform onto a plane as straight lines. This feature is extremely useful in navigation. However, the Mercator projection has the great disadvantage that regions at high latitudes (at, say, 50° or more) are considerably enlarged and distorted. Indeed, the poles are infinitely large in area.

We begin our studies involving the Mercator projection with a geometrical analysis of a small area on the surface of a sphere. The following mathematical relationships are obtained:

$$\frac{d\lambda}{d\phi} = \frac{\tan\theta}{\cos\phi} \quad \text{and} \quad \frac{ds}{d\phi} = \frac{R}{\cos\theta}, \tag{22.1}$$

where λ is the longitude, ϕ is the latitude, and θ is the angle that a certain path makes with a meridian. An incremental distance along the path is ds and R is the radius of the earth.

We rewrite the first of equations (22.1) in the form

$$\int_{\lambda_0}^{\lambda} \frac{d\lambda}{\tan\theta} = \int_{\phi_0}^{\phi} \frac{d\phi}{\cos\phi}. \tag{22.2}$$

The lower limits on the integrals indicate that when $\lambda = \lambda_0$, $\phi = \phi_0$. If we assume that θ is constant, that is, the path is always at the same angle with the meridians, then (22.2) can be integrated to give

$$\lambda - \lambda_0 = \tan\theta\left[\log_e \tan\left(\frac{\pi}{4} + \frac{\phi}{2}\right) - \log_e \tan\left(\frac{\pi}{4} + \frac{\phi_0}{2}\right)\right], \tag{22.3}$$

where the angles are expressed in radians. This equation is called the rhumb line or *loxodrome*. It transforms a straight line on the plane of a Mercator projection to a curved path on a sphere.

The second of equations (22.1) provides the expression

$$S = \frac{R}{\cos \theta}(\phi - \phi_0). \tag{22.4}$$

This simple equation gives the length S of the loxodrome between the point located at (λ_0, ϕ_0) and any other point (λ, ϕ).

In a moment, we return to our analysis of the Mercator projection and the Lambert azimuthal equal-area projection. But first, we need to take a close look at the following very remarkable phenomenon.

The Strange Behavior of the Mysterious Honking Bird

This is an appropriate place to describe and analyze the unbelievably strange behavior of the fascinating (and mythical) "honking bird." This extremely interesting winged creature is hatched on or very near the earth's equator. Then, when it reaches a certain age, it mysteriously begins an east-northeast-ward migration, which ultimately terminates at the North Pole, although a small fraction of these birds prefer to fly to the South Pole. Whether heading toward the north or the south, the honking bird nevertheless flies on a path that always holds a constant bearing with the meridians. Typically, the bearing angle θ is around 80°.

At this point we pause to let our mathematics catch up with us. To simplify our equation, we let $\phi_0 = 0$ in (22.3). Then we solve for ϕ to obtain

$$\phi = 2 \arctan e^{(\lambda - \lambda_0)/\tan \theta} - \pi/2. \tag{22.5}$$

This expression provides the value of ϕ (latitude) for any value of λ (longitude).

In addition, from (22.4), we have, with $\phi_0 = 0$,

$$S = R\phi/\cos \theta. \tag{22.6}$$

This expression gives the total length of the loxodrome, S, between the equator and any latitude ϕ.

Back to our honking bird: We select the value $\theta = 80°$ as the bird's "true heading" from north. The source of its flight is on the equator ($\phi_0 = 0$) near its well-known (but mythical) breeding ground along the northeastern shore of Lake Victoria in Africa ($\lambda_0 = 35°E$).

Remembering that ϕ and λ must be expressed in radians, we use equation (22.5) to compute the bird's position, $\phi = f(\lambda)$.

At this point it is suggested that you get your globe and world atlas. It will be helpful if you have them for the following analysis. With $\theta = 80°$, our honking bird is flying in an east-northeast direction. Hence, from (22.5), with $\lambda_0 = 35°E$, you easily calculate that when the bird is at longitude $\lambda = 125°E$ (i.e., one-quarter of the way around the world), it is at latitude $\phi = 16°N$. Your globe indicates that this location is east of the island of Luzon in the Philippines. Likewise, when the bird is at $\lambda = 215°E$ (i.e., $145°W$; halfway around the earth), then $\phi = 30°N$; this is northeast of Hawaii.

After one complete trip around the world, our honking bird is at latitude $\phi = 53°N$ (near Moscow) and after two complete revolutions, it is at latitude $\phi = 78°N$ (west of Franz Joseph Land in the Arctic Ocean).

A sketch of the bird's loxodromic path is shown in figure 22.3. We note that the radius of its path, measured from the earth's axis, continuously decreases as it moves northward. Indeed, when the bird gets extremely close to the North Pole, the radius of its path begins to approach zero. At this point, this otherwise entirely silent bird emits a very loud honking noise, which is a signal to himself to get out of the way. Because of this strange behavior—the bird's once-in-a-lifetime brief but extremely loud traffic blast—ornithologists have logically dubbed it the honking bird.

How far did our feathered friend go on his journey from Lake Victoria to the North Pole? With $\phi = \pi/2$ (i.e., $90°$) in equation (22.6), we obtain

$$S = \pi R / 2 \cos \theta. \tag{22.7}$$

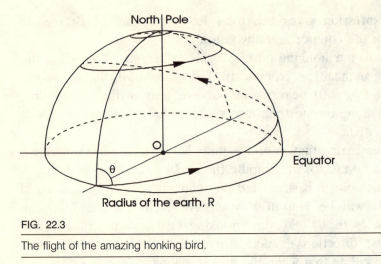

North Pole

Equator

θ

O

Radius of the earth, R

FIG. 22.3

The flight of the amazing honking bird.

With $\theta = 80°$ and $R = 6,370$ km, we compute that $S = 57,620$ km or about 35,810 miles.

The Mercator Projection Revisited

As shown in figure 22.1, the Mercator projection is constructed by wrapping a circular cylinder around the earth and tangent at the equator. If a tiny light source were at the center of a transparent globe, it would cast an image of the earth's land masses onto the cylinder. True enough, but this does *not* yield the Mercator projection; it might be called a gnomonic projection onto a cylinder. This projection is useless because there is extremely great distortion at high latitudes.

The Mercator projection is a conformal (i.e., angle-preserving) projection. From equation (22.1), the angle of an arbitrary path on a *spherical* surface is $\tan \theta = \cos \phi (d\lambda/d\phi)$. It is not hard to show that on a *cylindrical* surface, the angle of the path is $\tan \theta_* = R(d\lambda/dy)$, where y is the distance measured in the direction of the cylinder axis. Requiring that $\theta = \theta_*$ (i.e., equal-angle transformation), we obtain $dy = Rd\phi/\cos \phi$. In addition, we take $dx = Rd\lambda$, where x is the distance measured along the equator. Integrating these expressions for dx and dy gives

$$x = R\lambda \text{ and } y = R \log_e \tan\left(\frac{\pi}{4} + \frac{\phi}{2}\right). \tag{22.8}$$

These equations describe the positions of the meridians (lines of constant λ) and the parallels (lines of constant ϕ) on the projection. Clearly, on a Mercator map, the x-coordinate (east-west) is directly proportional to the longitude λ. However, the y-coordinate (north-south) is stretched increasingly as the latitude ϕ increases. When $\phi = \pi/2$, $y = \infty$.

Here are some examples involving the Mercator projection.

Part I

A jet aircraft flies from New York to Tokyo along the *great circle* route. How far is its journey? Show its path on a Mercator map.

From our atlas we obtain the following information: New York, longitude $\lambda_1 = 74°W$, latitude $\phi_1 = 41°N$; Tokyo, longitude $\lambda_2 = 140°E$, latitude $\phi_2 = 36°N$. The law of cosines for a spherical triangle gives the equation

$$\cos a = \sin \phi_1 \sin \phi_2 + \cos \phi_1 \cos \phi_2 \cos(\lambda_1 - \lambda_2). \tag{22.9}$$

The values of λ and ϕ indicated above are substituted into this expression to calculate $\cos a$. From this we obtain $a = 97°$, which is the angle of the great circle, measured at the earth's center, between New York and Tokyo. The distance S between the two cities is $S = (a/360)2\pi R$ and, with $R = 6,370$ km, we obtain $S = 10,785$ km or about 6,700 miles.

The path of the jetliner's flight is shown as curve A on the Mercator map of figure 22.4. After leaving New York, the airliner proceeds along the western shore of Hudson Bay and then across northern Canada. It reaches a maximum latitude of 70°N at a longitude of 146°W. This is close to Prudhoe Bay on the Arctic shore of Alaska—the hub of the North Slope's vast petroleum

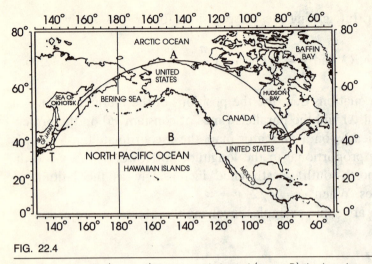

FIG. 22.4

Mercator map with (curve A) the great circle and (curve B) the loxodrome between New York and Tokyo.

fields. The flight then passes well north of the Bering Strait, flies through northeastern Siberia, along the Kamchatka Peninsula, and on to Tokyo.

Part II

On its return journey, our jet airplane flies from Tokyo to New York along the *loxodrome* route. Again, how far is its flight? Show its path on a Mercator map.

The values of λ and ϕ for New York and Tokyo are substituted into equation (22.3) to determine that the constant true heading of the jet's loxodrome path is $\theta = 87.5°$. Substituting back into (22.3) then provides the value of the longitude λ for any value of the latitude ϕ. This path is shown on curve B on the Mercator map of figure 22.4. This time our journey takes us across the wide expanse of the Pacific Ocean and over northern California, Denver, and Indianapolis, and then we land in New York.

From equation (22.4) we compute that the total length of our loxodromic flight is S = 12,745 km; this is about 18% longer than the great-circle flight.

The Lambert Azimuthal Equal-Area Projection

Another famous person involved in the development of cartography was the Swiss-German mathematician Johann Heinrich Lambert (1728–1777). Although he made contributions in many areas of mathematics, he will probably be best remembered for his work in cartography.

Perhaps his most noteworthy effort was what we now call the Lambert azimuthal equal-area projection, which he presented in 1772. This projection is employed extensively for maps of the polar regions.

We utilize this projection in figure 22.5 to display the northern hemisphere. The figure also shows the computed path of our jetliner flying from New York to Tokyo along the great circle (curve A) and also its route from Tokyo to New York along the loxodrome (curve B). Although this projection does not show true distances, it does indicate that the great circle is shorter than the loxodrome.

As the name implies, the Lambert azimuthal equal-area projection preserves areas. For example, the area of the island of Cuba (can you find it in figure 22.5? $\lambda = 80°W$, $\phi = 22°N$) is the same on the plane map as it is on a spherical globe. The undesirable but necessary distortion requires that Cuba be stretched in the east-west direction and shrunk in the north-south.

The Berghaus Star Projection

As mentioned, over the years a great many kinds of map projections have been developed. We conclude our introduction to cartography with an example involving a rather artistic map projection: the Berghaus star projection.

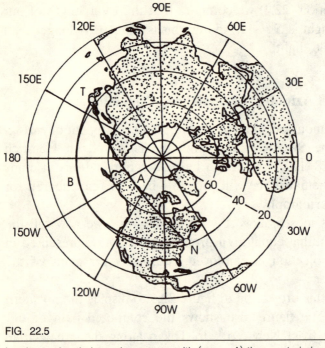

FIG. 22.5

Lambert azimuthal equal-area map with (curve *A*) the great circle and (curve *B*) the loxodrome between New York and Tokyo.

This projection was devised in 1879 by the German cartographer Heinrich Berghaus (1798–1884). It is shown in figure 22.6. In this map, the northern hemisphere is an azimuthal equidistant projection. The southern hemisphere consists of five triangular lobes. As Snyder and Voxland (1989) indicate, this projection is used mainly for artistic forms.

In chapter 2, we carry out an extensive analysis of the dimensions and geometrical features of five-pointed stars. Included in that analysis is the special case that appears in the Berghaus star projection.

One of the features of the Berghaus star is that its radius is twice the radius of the circle containing the northern hemisphere. This is the region defined by the equator passing through the five longitude locations shown in figure 22.6. From our earlier analysis, the internal angle at the points is $\alpha = 52.53°$.

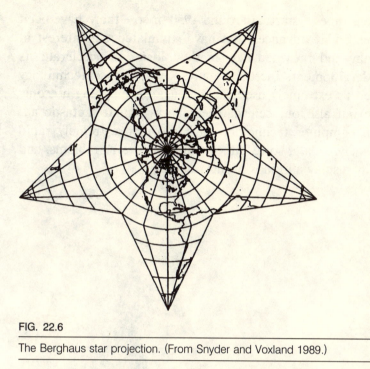

FIG. 22.6

The Berghaus star projection. (From Snyder and Voxland 1989.)

This is the shape of star that appears, or has appeared in the past, on the flags of various nations, including the flag of Vietnam during the period following World War II. It is interesting to note this rare linkage between vexillology (the science of flags) and cartography (the science of maps).

Looking Ahead in Cartography

Cartography is one of mankind's oldest sciences. It has been around as long as humans have navigated from here to there and as long as people have studied the skies. The science of cartography has evolved, slowly but steadily, over twenty-five to thirty centuries. During that very long period, many of the greatest mathematicians, astronomers, geographers, and navigators have contributed to the development of cartography and the making of maps.

In recent times—starting around 1950 or so—there have been many spectacular advances that have stimulated our interest in cartography and provided new methodologies to accelerate its further development. Incredible advances in space technology have given us extremely useful techniques for mapping not only the earth but also our celestial neighbors. Similar fantastic advances in computer technology have provided extremely rapid ways to analyze data, solve difficult mathematical problems, and graphically display all kinds of information.

23

Growth and Spreading and Mathematical Analogies

How fast does a plant or a person grow? What is the rate of increase of the population of a state or a nation? How quickly does a rumor or a disease spread through a certain community? How rapidly is a new technology adopted in a particular geographical setting? How fast does an innovation replace an established methodology?

To help us obtain answers to these and similar kinds of questions, we need to construct an appropriate mathematical framework. Such a framework is provided by the following simple differential equation:

$$\frac{dN}{dt} = aN\left(1 - \frac{N}{N_*}\right), \tag{23.1}$$

in which N is the magnitude of the growing or spreading quantity, t is the time, a is the growth or spreading coefficient, and N_* is the equilibrium value or carrying capacity.

The solution to this equation is given by the expression

$$N = \frac{N_*}{1 + (N_*/N_0 - 1)e^{-at}}, \tag{23.2}$$

where N_0 is the value of N at $t = 0$. This equation is called the Verhulst or logistic equation.

We note that if N is quite small compared to N_* (i.e., for "small" values of time t), then (23.2) reduces to the well-known Malthus or exponential equation, $N = N_0 \exp(at)$. For "large" values of t, the magnitude of N gradually approaches N_*. The logistic equation has been utilized for many years with great success in all kinds of problems involving growth and diffusion. An extensive analysis of the subject is given by Banks (1994).

It turns out that there are quite a few other relationships we could use, instead of the logistic equation, to describe the growth or spreading of something. For example, we could use the *normal probability* equation. Sometimes this is called the Gaussian equation, named after the great German mathematician Carl Friedrich Gauss (1777–1855).

Another expression we could employ for growth or spreading problems is the *arctangent-exponential* equation. It is defined by the differential equation

$$\frac{dN}{dt} = \frac{aN_*}{4} \sin\left(\frac{\pi}{N_*}N\right). \tag{23.3}$$

For the sake of a new adventure, let us see where this equation takes us.

First, we define the quantities $U = N/N_*$ and $T = at$, and substitute these into equations (23.1) and (23.3) to obtain the following two equations. The first one, of course, yields the solution (23.2).

$$\frac{dU}{dT} = U(1 - U) \text{ (logistic)}$$

$$\text{and } \frac{dU}{dT} = \frac{1}{4} \sin(\pi U) \text{ (arctan-exponential)}. \tag{23.4}$$

These expressions, now in so-called dimensionless form, are very similar. The first expression of (23.4) describes a parabolic curve and the second a sine curve. Both of them state that the growth rate $dU/dT = 0$ when $U = 0$ and $U = 1$ and also that $dU/dT = 1/4$ when $U = 1/2$. Figure 23.1 shows that these two curves are almost identical.

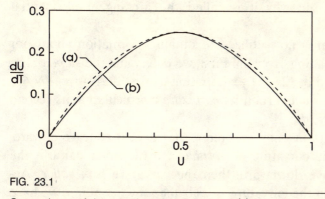

FIG. 23.1

Comparisons of the growth rate definitions of (a) the logistic equation and (b) the arctangent-exponential equation.

A number of years ago, the noted mathematician William Feller (1906–1970) made an extensive comparison of these two equations. Indeed, he made it a three-way comparison by including the normal probability equation. In his study, Feller (1940) analyzed the same sunflower growth data using the three different mathematical frameworks. He concluded that all three equations adequately describe growth phenomena and provide about the same answers.

With that as an introduction to this next topic, from here on we are concerned only with the arctangent-exponential relationship given by (23.3). We recast this equation in the form needed for integration:

$$\int_{N_0}^{N} \csc\left(\frac{\pi}{N_*}N\right) dN = \frac{aN_*}{4}\int_0^t dt, \tag{23.5}$$

where $\csc x = 1/\sin x$ is the cosecant. The lower limits of the integrals indicate that $N = N_0$ when $t = 0$. A table of integrals and a little algebra provide the final answer:

$$N = \frac{2N_*}{\pi}\arctan\left(\tan\frac{\pi N_0}{2N_*}e^{\pi at/4}\right). \tag{23.6}$$

Now you can see why it is called the arctangent-exponential equation.

This solution is now utilized to obtain information about our growth or spreading problem. First, we determine the *slope* of the curve expressed by (23.6) by computing its first derivative, dN/dt. Following this, we ascertain its *curvature* by calculating its second derivative, d^2N/dt^2.

If, over a certain range of time t, the curve is concave upward, we say that the curvature is *positive*. On the other hand, if the curve is concave downward, then the curvature is *negative*. Accordingly, when the curvature is *zero* then, by definition, we have identified what is called an *inflection point*.

Consequently, we simply set the second derivative equal to zero. This gives the value of the inflection-point time, t_i. Clearly, using this result in equation (23.6) gives the corresponding value N_i. Also, substituting the equation for t_i into the expression for the first derivative provides the maximum slope, $(dN/dt)_i$. If you want to understand all these procedures, you will confirm the following results:

$$t_i = \frac{4}{\pi a} \log_e \left(\cot \frac{\pi N_0}{2 N_*} \right); \; N_i = \frac{1}{2} N_*; \; \left(\frac{dN}{dt} \right)_i = \frac{1}{4} a N_*,$$

$$(23.7)$$

in which $\cot x = 1/\tan x$ is the cotangent.

Growth of the Population of California

Now for an example of an application. As most everyone knows, the population of California has increased rapidly during the past few decades. At the turn of the century, California ranked twenty-first among the states in population and in 1940 it ranked fifth. Now it is in first place by a very wide margin. California's population for the period from 1860 to 1990 is listed in table 23.1. It is displayed in graphical form in figures 23.2(a) and 23.2(b). For the moment, we ignore the solid curves in the figures.

TABLE 23.1

Growth of the population of California, 1860–1990

Year	t	N millions
1860	0	0.380
1870	10	0.560
1880	20	0.865
1890	30	1.214
1900	40	1.485
1910	50	2.738
1920	60	3.427
1930	70	5.677
1940	80	6.907
1950	90	10.586
1960	100	15.717
1970	110	19.971
1980	120	23.668
1990	130	29.126

Source: Wright (1992).

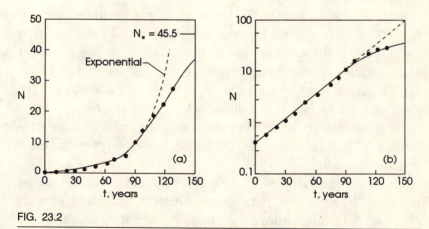

FIG. 23.2

Growth of the population of California. Population amounts are in millions. The year 1860 corresponds to $t = 0$. (a) Arithmetic and (b) semilogarithmic plots.

The pattern of the data points shown in figure 23.2(a) suggests that for small values of time t (and hence small N), the growth may be exponential. Let's try it. From (23.3) and recalling that for small x, $\sin x = x$, we obtain

$$\frac{dN}{dt} = \frac{\pi a}{4} N. \tag{23.8}$$

Integrating this equation gives

$$N = N_0 e^{\pi a t / 4}, \tag{23.9}$$

which is the exponential equation. Taking logarithms of (23.9),

$$\log_e N = \log N_0 + \frac{\pi a}{4} t. \tag{23.10}$$

This is an expression of the form $y = k_1 + k_2 x$, which is the equation of a straight line. So if we plot $\log_e N$ versus t, we should obtain a linear correlation with intercept $\log_e N_0$ and slope $\pi a / 4$. Determination of the intercept and slope gives the numerical values of N_0 and a.

Such a straight line is shown in the semilogarithmic plot of figure 23.2(b). It appears that the linear relationship extends from $t = 0$ (year 1860) to about $t = 100$ (year 1960). So we carry out a least-squares computation involving the first 11 entries of table 23.1. This yields the results $N_0 = 0.390$ and $a = 0.0469$. On this basis, we can say that California's population grew exponentially from 1860 to 1960. After that, crowding effects or growth-limitation factors—inherent in all logistic-type phenomena—began to be felt.

There are several ways to compute the value of the remaining quantity, N_*. For our purpose we simply indicate the answer: $N_* = 45.5$. In other words, the ultimate population of California will be about 45.5 million people.

So now we know that $N_0 = 0.390$, $a = 0.0469$, and $N_* = 45.5$. Substituting these numbers into equations (23.7) gives the inflection point $t_i = 117.0$ (year 1977), $N_i = 22.75$ million, and $(dN/dt)_i = 0.533$ million/year. Finally, substitution of these same numbers into (23.6) yields the solid lines shown in figures 23.2(a) and 23.2(b).

We conclude that the arctangent-exponential equation provides a good framework for quantitatively describing the growth of California's population.

Adoption of Hybrid Corn in Iowa

Here is another example. During the 1930s in midwest America, new strains of corn were developed and, because of their greater stamina and yields over open-pollinated varieties, began to be utilized by farmers in the region.

This hybrid corn innovation began in the heart of the Corn Belt—the states of Iowa and Illinois—where its adoption by farmers was slow at first, then more rapid, and finally leveled off after virtually all farmers in the region had switched to the new variety.

An extensive study of this adoption of hybrid corn was carried out by Griliches (1960) covering the period from 1933 to 1958. For each of 31 states, mostly in the midwest and south, he determined the percentage of acreage planted with hybrid corn to the total corn acreage available. We shall examine and analyze what happened in the state of Iowa.

The results obtained by Griliches for Iowa from 1933 to 1942 are listed in table 23.2 and plotted in figures 23.3(a) and 23.3(b).

Our analysis of hybrid corn adoption in Iowa is similar to the one we carried out for the population of California. The results are $U_0 = N_0/N_* = 0.0096$ and $a = 1.12$. Our problem of hybrid corn is easier to analyze than the problem of California's population because, from Griliches' data, we already know that $N_* = 100\%$.

Substitution of these numerical values of U_0 and a into equations (23.7) provides the inflection point: $t_i = 4.76$ (year 1937.76), $U_i = 0.50$, and $(dU/dt)_i = 0.28$ per year. Substituting these same numbers into (23.6) gives the solid lines of figures 23.3(a) and 23.3(b).

TABLE 23.2

Adoption of hybrid corn in Iowa, 1933–1942
Amounts are ratios of hybrid corn acreage to total corn acreage.

Year	t	$U = N / N_*$
1933	0	0.01
1934	1	0.02
1935	2	0.05
1936	3	0.15
1937	4	0.30
1938	5	0.52
1939	6	0.73
1940	7	0.90
1941	8	0.97
1942	9	0.98

Source: Griliches (1960).

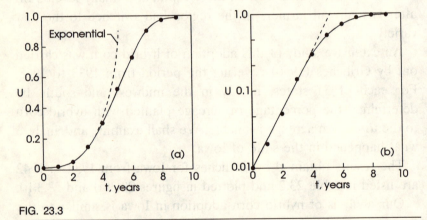

FIG. 23.3

Adoption of hybrid corn in Iowa. The year 1933 corresponds to $t = 0$. (a) Arithmetic and (b) semilogarithmic plots.

Again, we can say that the arctangent-exponential equation describes the adoption of hybrid corn in Iowa rather nicely.

The Loxodrome Curve on a Sphere

We now shift the time origin of equation (23.6) to the right by an amount equal to the inflection point time, t_i. In other words,

$t = 0$ now passes through the inflection point. Consequently, (23.6) becomes

$$N = \frac{2N_*}{\pi}\arctan(e^{\pi at/4}). \tag{23.11}$$

This is a remarkable answer! Does this equation look familiar? It should; it is essentially the same equation we obtained for the loxodrome curve in chapter 22. Recall that the loxodrome is a curve that maintains a constant angle with the meridians on a sphere. On a Mercator projection it appears as a straight line.

In that chapter we determined that the equation of the loxodrome passing through the point $\lambda = 0$, $\phi = 0$ on a sphere (where λ and ϕ are the longitude and latitude) is

$$\phi + \frac{\pi}{2} = 2\arctan(e^{\lambda/\tan\theta}), \tag{23.12}$$

in which θ is the constant angle between the curve and the meridians.

This is a very interesting result. Comparison of equations (23.11) and (23.12) provides the following list of analogous quantities:

Growth or adoption curve	Loxodrome curve
N	$\phi + \pi/2$
t	λ
N_*/π	1
$\pi a/4$	$\cot\theta$

This analogy says that the growth of the population of California and the adoption of hybrid corn in Iowa are described by the same mathematics as the loxodrome curve on the surface of a sphere.

In other words, the latitude ϕ, measured from the South Pole, of a point on a loxodrome curve on a sphere corresponds exactly to the magnitude of a growing quantity (e.g., the population of

California) and to the amount of new technology adoption (e.g., hybrid corn in Iowa). The longitude λ of the point on the loxodrome corresponds to the growth or adoption time.

On second thought, the correspondence or equivalence of the quantities is not that remarkable; it is really more a fortunate coincidence. We arbitrarily selected the *arctangent-exponential* for the growth-adoption problems and so we picked up the loxodrome analogy. Had we stayed with the *logistic* equation the analogy would not have appeared.

Nevertheless, the above result does bring out something that is very important; the recognition and utilization of *mathematical analogies*. Such analogies provide extremely useful methodologies for the applied mathematician, engineer, and scientist.

There Are All Kinds of Mathematical Analogies

Since there is an almost endless list of mathematical analogies, we shall look at only three basic kinds. However, even these three provide a large number of interesting and very helpful applications. An excellent reference on analogies is the book by Soroka (1954).

Mechanical and Electrical Circuit Analogies

The mathematical basis for these analogies is Newton's second law (for mechanical circuits) and Ohm's and Kirchhoff's laws (for electrical circuits).

The mechanical circuit features a system comprised of a mass (m), a damping chamber (c), and a spring (k). The displacement (x) and velocity (v) of the oscillating system occur as the result of an external force (F).

With precisely the same wording, the electrical circuit features a system comprised of an inductance (L), a resistance (R), and a capacitance (C). The electrical charge (q) and current (i) in the oscillating system occur as a result of an electrical voltage (E).

These mechanical and electrical circuits find exact mathematical analogies in structural engineering (oscillation of suspension bridges), naval architecture (pitching and rolling of ships), biology (prey-predator phenomena), economics (supply, demand, and investment functions), and numerous other disciplines. Currently, there is a good deal of interest in these topics in connection with nonlinear dynamics and chaos theory. MacDonald (1989) discusses the problem of "damped linear oscillators with a delayed restoring force." Thompson and Stewart (1986) consider phenomena involving "forced nonlinear oscillators with periodic and strange attractors."

Soap Film and Membrane Analogies

There are many very useful applications of analogies involving soap films and thin membranes.

For example, suppose that a soap film or flexible membrane covers the horizontal top of an otherwise open conduit of arbitrary cross-section. A slight air pressure beneath the film raises the film upward except along the conduit boundaries. The deflection of the film above the original datum is measured. Though it is far from obvious in our brief consideration, what has been achieved with these measurements is a solution to the Laplace equation. This equation is extremely important in many problems of electrostatics, elasticity, heat conduction, porous media flow, and hydro- and aerodynamics.

Another application of the soap film analogy relates to the branch of mathematics called the calculus of variations. As mentioned in an earlier chapter, this subject is concerned with problems involving minimum length or minimum area or minimum energy. For example, it can be shown that the "catenary of revolution" is the minimum area of a soap film connecting two coaxial circular hoops of different diameters. A good deal of useful information concerning soap bubbles and soap films is given by Isenberg (1992). Soap films are also discussed by Kappraff (1990) in connection with the problem of minimal total lengths of highway networks.

Laminar and Potential Flow Analogies

Over the years, many investigators have employed electrolytic tanks and electrically conducting paper to solve all kinds of problems dealing with so-called potential flow phenomena. A potential flow is one in which the velocity is directly proportional to the gradient of the potential (e.g., pressure).

Briefly, in an electrolytic tank (or on conducting paper), electrodes are placed along specified portions of the boundary of a shallow tank containing an electrolyte (e.g., salt water). An object (e.g., an airfoil cross-section) made of nonconducting material is placed in the center of the tank. A certain voltage is applied to one electrode; the other electrode is grounded.

A probe is then employed to measure the voltage at any point in the tank, especially near the object being studied. Connection of the points of equal voltage determines the positions of the equipotentials. Appropriate changes of the locations of the boundary electrodes and another round of voltage measurements establishes the locations of the streamlines. In this way, the entire flow field around the object is determined.

Another very useful mathematical analogy is provided by the Hele–Shaw cell. Although this analog has been known for a long time, there is considerable renewed interest in this methodology because of current interest in the subject of ground water recharge of aquifers and secondary oil recovery in petroleum reservoirs.

A Hele–Shaw cell consists of two glass plates separated by a very small gap. A highly viscous fluid occupies the gap space. Indeed, two or even three immiscible fluids, of different densities and viscosities, can be employed to provide solutions to many kinds of difficult problems. When the fluids are given different colors, the patterns created yield spectacular dynamic art.

As in the case of electrolytic tanks, a test object (e.g., the sheet-piling configuration under a dam) is installed in the cell and appropriate boundary conditions are applied along the edges.

Colored dye streams show the flow patterns. With this information, velocity and pressure distributions can be computed.

Current renewed interest in Hele–Shaw cells is due to the fact that there is exact analogy between Poiseuille's law (which governs the behavior of laminar flow between closely spaced plates) and Darcy's law (which governs the behavior of flow in porous media). This analogy has been known for a great many years. However, in recent times, applied mathematicians and petroleum engineers have become interested in problems of instability of the interface between a driving fluid and a driven fluid as they flow through the cell. Such instabilities produce complex patterns of viscous fingering and fractal structures. This is an important consideration in problems of effective oil recovery in petroleum reservoirs.

An easy to understand article about Hele–Shaw cells is provided by Walker (1987); a not-so-easy one is given by Schwartz (1988).

24

How Long Is the Seam on a Baseball?

Or, if you prefer, how long is the groove on a tennis ball? Now we all know that (a) this is not one of the great unsolved problems of mathematics and (b) you will enjoy the World Series or Wimbledon just as much never knowing. However, if you welcome an opportunity to deflate the egos of certain friends or relatives who, by their own admission, know practically everything there is to know about baseball or tennis, your moment has arrived.

We have long been aware of the fact that the seam on a baseball, the groove on a tennis ball, and the dimples on a golf ball, enormously affect the aerodynamics of ball flight. So a study of these geometrical features of sporting balls certainly relates to the very interesting problem of ball trajectories.

In the following sections, we examine a number of relationships of spherical geometry. For these endeavors, it may be helpful if you have the following items: a single-color plastic ball six to eight inches in diameter, some marker pens (two or three colors, easily erasable), a centimeter scale, and some string.

Neatly glue pieces of string onto the ball to establish an equator and four meridians spaced at 90°. Mark the poles N and S. Measure the circumference of the ball, C, or diameter, D, and compute the radius, R.

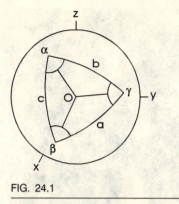

FIG. 24.1

Definition sketch for a spherical triangle.

The mathematics of our baseball seam problem is not very difficult. Mostly, it involves only spherical trigonometry; later on we use a bit of differential calculus to compute some slopes.

The following two fundamental laws are utilized in our analysis; the various symbols are defined in figure 24.1.

The *law of cosines* states that

$$\cos a = \cos b \cos c + \sin b \sin c \cos \alpha, \qquad (24.1)$$

and two similar equations for $\cos b$ and $\cos c$.

The *law of sines* states that

$$\frac{\sin a}{\sin \alpha} = \frac{\sin b}{\sin \beta} = \frac{\sin c}{\sin \gamma}. \qquad (24.2)$$

A photograph of a baseball, a tennis ball, and a golf ball is shown in figure 24.2. Some of the physical features of the three types of balls are listed in table 24.1.

It will be helpful if you have, in addition to the plastic ball mentioned above, a baseball, a tennis ball, and perhaps a golf ball. You might even want to acquire a basketball.

Now a word about the all-important quantity designated "minimum latitude, ϕ_0" in table 24.1. With your centimeter ruler or scale, carefully measure, on your baseball, the distance between the seams at the point where the seams come closest

FIG. 24.2

Sporting balls: base, tennis, and golf.

TABLE 24.1

Baseballs, tennis balls, and golf balls

	Radius R, centimeters	Mass m, grams	Minimum latitude ϕ_0, degrees
Baseball	3.68	145	25
Tennis ball	3.20	58	40
Golf ball	2.10	46	—

together. You should get about 3.2 centimeters. Next, measure the entire circumference of the baseball; the result is approximately 23.0 centimeters. Since there are 360° along the circumference of the ball, we define the arc angle $2\phi_0 = (3.2/23.0)$ $(360) = 50.08°$. Therefore, $\phi_0 = 25.04°$, which we round off to $\phi_0 = 25°$. Similar measurements on a tennis ball give $\phi_0 = 40°$.

In the photograph of figure 24.2, the North Pole is at the top and the South Pole at the bottom. The minimum seam gap and the minimum latitude you just now determined is at the left edge

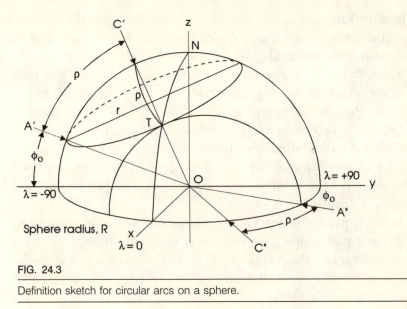

FIG. 24.3

Definition sketch for circular arcs on a sphere.

of the ball. This corresponds to longitude $\lambda = -90°$. A vertical line (not shown) halfway across the ball defines the meridian, $\lambda = 0$. The right edge is $\lambda = +90°$.

The equator, of course, is the great circle whose plane is perpendicular to the polar axis. North and south latitudes, ϕ, are measured from the equator along great circles passing through the poles. These are meridians. In case you have forgotten, a great circle is one whose plane passes through the sphere's center.

We Begin the Analysis of the Problem

The definition sketch for our problem is shown in figure 24.3. The North Pole, N, is at the top. Point A', the one you just determined, is located at $\lambda = -90$, $\phi = \phi_0$.

The following is very important. With baseball in hand and oriented as in figure 24.2, convince yourself that the extreme right-hand point of the seam is located at $\lambda = 90 - \phi_0$, $\phi = 0$. This point is labeled A'' in figure 24.3. Equally important is the fact that the seam must pass through the point $\lambda = 0$, $\phi = 45$.

These two important features are required because the remaining portion of the baseball's entire template must be identical to the portion we are analyzing.

Here we go. With attention to figure 24.3, we are going to pass a circle—though not a great circle—through points A' and T and another circle through points T and A''; T is the meeting point of the two circular arcs comprising the seam.

The arc radius of each of these two "small circles" is ρ. The center of one is located at $\lambda = -90$, $\phi = \phi_0 + \rho$; the center of the other is at $\lambda = 90 - (\phi_0 + \rho)$, $\phi = 0$. The two centers are labeled C' and C''.

A short pause follows. On the plastic ball you have elaborated with an equator and four meridians made of string, label with your marker pen the points A', T, and A''. Why not use $\phi_0 = 25$, because that's a baseball. Since you know the circumference of your own plastic ball, you can easily calculate the arc distance corresponding to $25°$.

Now we look at the spherical triangle $C'NT$, where N is the North Pole. From the law of cosines, equation (24.1), we have

$$\cos \rho = \cos[90 - (\phi_0 + \rho)] \cos 45$$
$$+ \sin[90 - (\phi_0 + \rho)] \sin 45 \cos 90. \qquad (24.3)$$

By the way, in our analyses, we will frequently be using the following relationships:

$$\sin(x \pm y) = \sin x \cos y \pm \cos x \sin y,$$
$$\cos(x \pm y) = \cos x \cos y \mp \sin x \sin y. \qquad (24.4)$$

Remember also that $\sin 0 = 0$, $\sin 90 = 1$, $\cos 0 = 1$, and $\cos 90 = 0$.

You should demonstrate, from equation (24.3), that the formula to compute the arc radius of the circle centered at C' is

$$\rho = \arctan(\sqrt{2} \sec \phi_0 - \tan \phi_0), \qquad (24.5)$$

where $\sec x = 1/\cos x$ and $\tan x = \sin x/\cos x$. Substituting

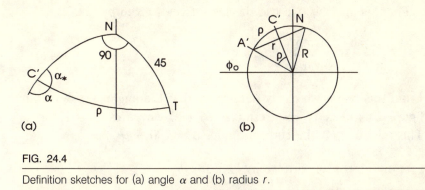

FIG. 24.4

Definition sketches for (a) angle α and (b) radius r.

$\phi_0 = 25$ into (24.5) gives $\rho = 47.57$. This establishes the location of point C'.

A similar analysis involving spherical triangle $C''NT$ again yields (24.5) and $\rho = 47.57$; this locates points C''. Now you can show points C' and C'' on your plastic ball. Point C' is at $\lambda = -90$, $\phi = 72.57$, and point C'' is at $\lambda = 17.43$, $\phi = 0$.

Next, as accurately as you can, use your scale to draw circles centered at C' and C'', with arc radius $\rho = 47.57$. This is easy. Just compute for your plastic ball the arc length corresponding to 47.57°. With your marker pen, locate 10 or 12 points measured from C' and from C'' and connect these points to make the two circles.

What you have constructed should look like figure 24.3 but better, because you are in three dimensions. If you have been reasonably careful, the two circles should meet at point T. An interesting result: The curve $A'TA''$ is one-fourth of the baseball seam. Show it on your plastic ball with a different colored marker pen.

We Continue the Analysis of the Problem

As the next step in our analysis, we again look at spherical triangle $C'NT$. This time we have to analyze it twice; the first time for the case in which $\phi_0 < 45$ and the second for $\phi_0 > 45$. A definition sketch for the first case is provided in figure 24.4(a).

For this analysis, we use the law of sines given by (24.2):

$$\frac{\sin \rho}{\sin 90} = \frac{\sin 45}{\sin \alpha_*}. \tag{24.6}$$

From this result we easily determine that the angle α_* shown in figure 24.4(a) is $\alpha_* = \arcsin(\csc \rho / \sqrt{2})$ where $\csc x = 1/\sin x$. This tells us that the central angle, $\alpha = 180 - \alpha_*$, is

$$\alpha = 180 - \arcsin(\csc \rho/\sqrt{2}), \text{ for } \phi_0 < 45. \tag{24.7}$$

For the case $\phi_0 > 45$, again using spherical triangle $C'NT$, we get

$$\alpha = \arcsin(\csc \rho/\sqrt{2}), \text{ for } \phi_0 > 45. \tag{24.8}$$

We note that the arc length from A' to T is

$$S_1 = 2\pi r\left(\frac{\alpha}{360}\right), \tag{24.9}$$

in which, as seen in figure 24.4(b), $r = R \sin \rho$. This equation says that the length of the circular arc from A' to T is simply the proportion, $\alpha/360$, of the circumference of a plane circle of radius r, where $r = R \sin \rho$.

Since $360°$ is equal to 2π radians, equation (24.9) gives $S_1 = R \alpha \sin \rho$. This is the length from A' to T. By symmetry, it is also the length from T to A''. So the total length from A' to A'' is $S_2 = 2S_1 = 2R\alpha \sin \rho$. This is the length of the curve drawn on your plastic ball. Since this is one-fourth of the total, the length of the entire seam is

$$S = 8R\alpha \sin \rho. \tag{24.10}$$

We Obtain the Answer to the Problem

We established that $\phi_0 = 25$ for a baseball and, from equation (24.5), we computed that $\rho = 47.57$. Therefore, $\sin \rho = 0.7381$. Also, from (24.7), the central angle is $\alpha = 106.67° =$

1.8617 radians. Substituting these numbers into (24.10) gives $S/R = 10.994$.

As indicated in table 24.1, the radius of a baseball is $R = 3.68$ cm. So we finally determine that the length of the seam on a baseball is 40.46 centimeters or 15.93 inches.

Please check these results by direct measurements on your plastic ball and your baseball. Since there are 216 stitches on a baseball, how many stitches per inch do you determine (a) by counting and (b) by computing?

QUESTION FOR YOUR COFFEE BREAK You own a smart ant that always runs at a speed of 1.0 centimeters per second. How long will it take the ant to cover the length of a tennis ball groove assuming no stops en route?

Answer. Almost exactly 30 seconds.

WHAT IS THE ANGLE BETWEEN THE TWO PLANES? On the way back from coffee break, you should stop at the market and buy either a nice round ball of Dutch cheese or a medium-size cantaloupe. On one or the other, please reproduce the art work you have created on your plastic ball.

Next, with your knife or small saw, carefully cut the cheese (or cantaloupe) through the planes defined by the circles centered at C' and C''. Then place two pieces of stiff cardboard against the flat surfaces; the edges of the cardboard will meet along a line passing through point T. You will observe a certain angle between the cardboard planes. Measure this angle with your scale or protractor.

If you have used the value $\phi_0 = 25°$ throughout and if you have carried out the slicing and measuring operations with reasonable care, you will determine that the angle between the planes is about 85°.

To solve the problem mathematically gets us into the topic of solid analytic geometry. The problem is pretty easy. It starts with recognizing that the angle between the two planes is the same as the angle between the two lines perpendicular to the planes. Call this angle σ. It can be computed from the relationship

$$\cos \sigma = l_1 l_2 + m_1 m_2 + n_1 n_2, \qquad (24.11)$$

TABLE 24.2

Summary of some main results
All quantities are in degrees except the dimensionless
curve length, S/R

ϕ_0 minimum latitude	ρ small-circle radius	α central angle	σ angle between planes	S/R curve length	Remarks
0	54.74	120.00	70.53	13.68	
10	51.56	115.47	76.89	12.63	
20	48.77	109.91	82.47	11.54	
25	47.57	106.67	84.85	10.99	baseball
30	46.55	103.08	86.90	10.45	
40	45.20	94.76	89.60	9.39	tennis ball
45	45.00	90.00	90.00	8.89	
50	45.24	84.82	90.48	8.41	
60	47.63	73.16	95.26	7.55	
70	54.22	60.65	108.44	6.87	
80	67.98	49.71	135.96	6.43	
90	90.00	45.00	180.00	6.28	basketball

where l, m, and n are the so-called direction cosines of the two lines, one passing through C' (subscript 1) and the other through C'' (subscript 2).

In our problem, it is easy to compute the direction cosines because C' is located along $\lambda = -90$ and C'' along $\phi = 0$. Without going into details, we obtain the answer

$$\sigma = \arccos[\cos^2(\phi_0 + \rho)]. \tag{24.12}$$

If $\phi_0 = 25$ (baseball), we get $\sigma = 84.85$; if $\phi_0 = 40$ (tennis ball), then $\sigma = 89.60$, and if $\phi_0 = 45$ then $\sigma = 90$. If $\phi_0 = 90$ then $\sigma = 180$ or $\sigma = 0$, if you prefer. In this limiting case, the two planes coincide; this is rather obvious.

Summary of Results

Before we go to a final topic or two, we summarize the main results obtained so far; this information is listed in table 24.2.

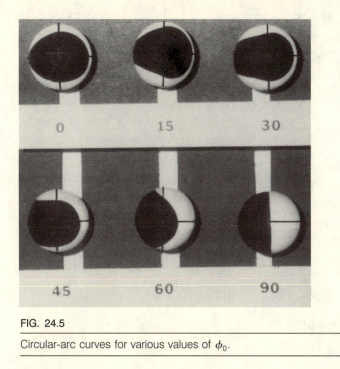

FIG. 24.5

Circular-arc curves for various values of ϕ_0.

As a visual aid, some painted ping-pong balls are shown in figure 24.5. The six balls show the circular arc curves (i.e., seams or grooves) for six values of ϕ_0. A wonderment: Why did they select $\phi_0 = 25°$ for a baseball and 40° for a tennis ball?

If I Know the Longitude then What Is the Latitude?

The question is rephrased. We have a large zeppelin we are ordered to fly from point A', $\lambda = -90$, $\phi = \phi_0$, to point A'', $\lambda = 90 - \phi_0$, $\phi = 0$. For reasons we are not permitted to know, we are instructed to follow a baseball-seam-type curve in our journey from A' to A''.

We have state-of-the-art navigational equipment aboard and so we know our position (λ, ϕ) at all times. Here is the question: Knowing our longitude λ what should be our latitude ϕ in order to be on course. In other words, what is the equation of the curve $\phi = f(\lambda)$ between A' and A''?

It is necessary to solve the problem in two parts: (a) for $\lambda < 0$ (i.e., in the western hemisphere) and (b) for $\lambda > 0$ (the eastern hemisphere). The Greenwich meridian is $\lambda = 0$. We are not going to go through all the details. However, you might want to do this as a homework problem.

Region $\lambda < 0$

On your plastic ball, with the nice curve corresponding to $\phi_0 = 25$, select any point on the curve midway between A' and T. Label this point P' and draw, through P', the arc of a great circle from the North Pole to the equator.

Using the spherical triangle $C'NP'$ and the law of cosines, equation (24.1), we write

$$\cos \rho = \cos[90 - (\phi_0 + \rho)]\cos(90 - \phi)$$
$$+ \sin[90 - (\phi_0 + \rho)]\sin(90 - \phi)\cos[-90 - (-\lambda)].$$
$$(24.13)$$

Utilizing equations (24.4) several times, employing (24.5) to express ρ in terms of ϕ_0, and going through some algebra—including the solution of a quadratic equation—we obtain the answer:

$$\phi = \arccos\left[\frac{q \sin \lambda}{2 + q^2 \sin^2 \lambda}\left(1 + \frac{\sqrt{2}}{q} \csc \lambda\sqrt{1 + q^2 \sin^2 \lambda}\right)\right],$$
$$(24.14)$$

in which

$$q = \frac{1 - \sqrt{2} \sin \phi_0}{\cos \phi_0}.$$
$$(24.15)$$

This equation is a little messy but you can easily handle it on your calculator. It is important to remember that (24.14) refers to the region *west* of $\lambda = 0$.

Region $\lambda > 0$

On your plastic ball, select another point, P'', on the curve somewhere between T and A''. This time, using spherical triangle $C'' NP''$ and the law of cosines, we get

$$\phi = \arccos\left[\frac{\cos \rho}{\sin(\phi_0 + \rho)\cos \lambda + \cos(\phi_0 + \rho)\sin \lambda}\right],$$

(24.16)

where, as in equation (24.5),

$$\rho = \arctan\left(\sqrt{2} \sec \phi_0 - \tan \phi_0\right).$$

(24.17)

Maybe you can find a neater way to express this answer. Again, remember that (24.16) is for the region *east* of $\lambda = 0$.

Well, with equations (24.14) and (24.16), we have what we want: the equation of the curve $\phi = f(\lambda)$. You might feel like computing this curve with $\phi_0 = 25$ and comparing the results with the curve you have already plotted on your plastic ball.

Here's another topic: You will observe on your plastic ball that in the $\lambda > 0$ region, the curve reaches a maximum latitude ϕ_m. This is a very interesting result. Using differential calculus, you can show, from (24.16), that this maximum point has the coordinates

$$\lambda_m = \arctan\left[\cot(\phi_0 + \rho)\right]; \quad \phi_m = \rho.$$

(24.18)

Remember that this result is obtained from equation (24.16), which is valid only for $\lambda > 0$. If $\phi_0 > 45$, there are no maxima in the eastern hemisphere ($\lambda > 0$) but there are in the western ($\lambda < 0$). We can determine those maxima by inspection: $\lambda_m = -90$, $\phi_m = \phi_0$. Take a look at the ping-pong balls in figure 24.5.

One more thing needs to be established. We would expect that at $\lambda = 0$, the value of ϕ should be the same whether it is computed from (24.14) (for $\lambda < 0$) or (24.16) (for $\lambda > 0$). This is indeed the case, as you might want to confirm. Regardless of the value of ϕ_0, the curve always passes through point T: $\lambda = 0$, $\phi = 45$.

Furthermore, at point T we want the slope of the curve, $d\phi/d\lambda$, computed from (24.14) to be the same as the slope computed from (24.16). In other words (back to our zeppelin flight), we do not want our ancient airship to undergo any sudden drastic course change as it sails over $\lambda = 0$, $\phi = 45$—a point 50 kilometers east-northeast of Bordeaux, France.

As we know by now, computing the slope of a curve means taking its first derivative. You have to be rather brave to differentiate equation (24.16) (for $\lambda > 0$) and almost foolhardy to try to take on equation (24.14) (for $\lambda < 0$).

This is clear enough. However, we do not need to know the slope at *every* point along the curve—only the slope near $\lambda = 0$. So what we shall do is "linearize" equations (24.14) and (24.16). This means that for small λ we can use the approximations $\sin \lambda = \lambda$ and $\cos \lambda = 1$.

Doing this to (24.14) yields the simple expression

$$\phi = \arccos\left(\frac{q}{2}\lambda + \frac{1}{\sqrt{2}}\right), \tag{24.19}$$

where, as before, $q = (1 - \sqrt{2} \sin \phi_0)/\cos \phi_0$. Taking the first derivative of this equation and then substituting $\lambda = 0$ gives

$$\frac{d\phi}{d\lambda} = \frac{1 - \sqrt{2} \sin \phi_0}{\sqrt{2} \cos \phi_0} \text{ at } \lambda = 0. \tag{24.20}$$

Linearizing equation (24.16) provides the relationship

$$\phi = \arccos\left(\frac{\cos \rho}{a + b\lambda}\right), \tag{24.21}$$

where $a = \sin(\phi_0 + \rho)$ and $b = \cos(\phi_0 + \rho)$. Differentiating this equation and substituting $\lambda = 0$ again yields (24.20), as we would hope. Therefore the curve $\phi = f(\lambda)$ is *continuous* at $\lambda = 0$, that is, there is no abrupt angle change.

TABLE 24.3

Another summary of some main results
Units are degrees.

ϕ_0 minimum latitude	λ_m longitude of maximum point	$\phi_m = \rho$ latitude of maximum point	θ meridian angle at $\lambda = 0$	Remarks
0	35.26	54.74	45.00	
10	28.44	51.56	52.54	
20	21.23	48.77	61.21	
25	17.43	47.57	66.06	baseball
30	13.45	46.55	71.31	
40	4.80	45.20	83.23	tennis ball
45	0 to −90	45.00	90.00	
50	−90	50.00	97.39	
60	−90	60.00	114.20	
70	−90	70.00	133.88	
80	−90	80.00	156.15	
90	−90	90.00	180.00	basketball

To be precise, on a sphere the slope of a curve is given by the equation

$$\frac{1}{\cos\phi}\frac{d\phi}{d\lambda} = \frac{1}{\tan\theta},$$

(24.22)

where θ is the angle between the curve and the meridian. This expression was developed in chapter 22, in which we computed the curve called the rhumb line or loxodrome.

Well, since equation (24.20) gives us the value of $d\phi/d\lambda$ at point T and since $\phi = 45$ at point T, we can easily calculate the meridian angle θ at T for any value of the basic parameter ϕ_0. Table 24.3 lists these values of θ along with the maximum point coordinates.

PROBLEM We return to our zeppelin and its proposed flight from $\lambda = -90$, $\phi = \phi_0$ to $\lambda = 90 - \phi_0$, $\phi = 0$. Select any value of ϕ_0 you like except $\phi_0 = 90$ (basketball) — there are no facilities for handling zeppelins at the North Pole. Calculate the zeppelin's path, $\phi = f(\lambda)$, and plot it on your plastic ball. Determine the length of its journey; the radius of the earth is $R = 6,370$ km.

From your atlas, find out where the journey of the zeppelin begins and ends. If its velocity $V = 100$ km / hr, how many days does the journey take? Where in the world is the point of maximum latitude? As the zeppelin crosses the point $\lambda = 0$, $\phi = 45$ (near Bordeaux, France), what is its heading from true north?

In this chapter, we have carried out analyses on a number of fairly complicated problems involving spherical coordinates. The excellent book by Gellert et al. (1977) covers the subjects of spherical geometry and spherical trigonometry in a manner that is easy to understand.

Although we are not yet through with the problem of the baseball seam, this is a good place to pause. In the next chapter, we continue with the same general subject but with applications to topics that have nothing whatsoever to do with baseballs.

Sincerest appreciation is expressed to the Wilson Sporting Goods Company for providing very helpful information concerning the geometry of baseball templates.

25

Baseball Seams, Pipe Connections, and World Travels

We have established that the length of the seam on a baseball is 10.99 times the radius of the ball and that the length of the groove on a tennis ball is 9.39 times its radius. Very interesting. What else can we get out of our lengthy analysis?

This is a good question. We are now going to look at two problems that have nothing to do with a baseball but everything to do with the mathematics we developed in analyzing a baseball. The first problem is quite a practical one in civil engineering; the second problem deals with some aspects of cartography and geography.

A Nice Way to Connect Pipes

Here's the engineering problem. We have a deep well (or indeed an array of wells) penetrating a ground water aquifer or a petroleum reservoir. In the case of the aquifer, we plan to pump water into an injection well for ground water recharging, or perhaps to build a barrier to eliminate salinity intrusion from the ocean.

In the case of the petroleum reservoir, we intend to inject water into an oil-bearing layer to force the oil to move to nearby wells, where it is pumped out. When the reservoir was pumped the first time, it was not possible to remove all the oil, so this "secondary oil recovery" technique of injecting a "water flood" is used.

Some distance from the top of the injection well, we have a horizontal pipe that will deliver water to the well head. In the case of ground water recharging, the water in the delivery pipe is probably the effluent from a waste treatment plant. In the case of secondary oil recovery, the water is being delivered from a process tank in which chemicals are added to reduce the surface tension between the sand particles and the oil, so that more oil can be extracted.

In either case, the engineers have intentionally left a suitable distance between the end of the elevated horizontal water delivery pipe and the well head located at ground level. There are good reasons for leaving this gap. The engineers want to install a curved connection pipe between the two points (a) to allow for significant expansion and contraction of the steel pipe due to temperature changes and (b) to provide for a smooth transition of the flow of water from the horizontal delivery pipe to the vertical injection wall. If the connection were simply a 90° elbow at the well head, friction losses and hence pumping costs would be substantially increased.

With that introduction we get to the problem. Our approach closely follows that given by Kosko (1987) in his analysis of "three-dimensional pipe joining." A definition sketch for our problem is shown in figure 25.1.

The surveyors in the field measure and report the following two quantities to the engineering design office: (a) the height h of the horizontal delivery pipe and (b) the horizontal distance L between the end of the delivery pipe and the well head. With this information concerning the numerical values of h and L, the design engineers then compute the minimum latitude ϕ_0 and radius R of an equivalent but entirely fictitious sphere. The remainder of the problem is easy. Using the same symbols, the equations we derived in the previous chapter are employed to

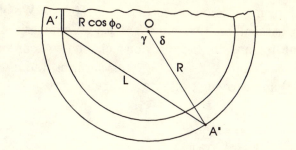

FIG. 25.1

Definition sketch for the pipe connection problem. A': water delivery pipe (elevation h), A'': well head (elevation 0).

compute

The length of the two circular arcs of the connection pipe between A' and A'': S

The bend radius: r

The bend angle: α

The angle between the planes of the two arcs: σ

The location of the maximum height: h_m

We now back up. As noted, the only information obtained from the field is the vertical height h and the horizontal distance L. How can ϕ_0 and R be determined with only this information? To answer this, we first employ the plane law of cosines,

$$c^2 = a^2 + b^2 - 2ab \cos \gamma, \tag{25.1}$$

which, with regard to figure 25.1, takes on the form

$$L^2 = (R \cos \phi_0)^2 + R^2 - 2(R \cos \phi_0)(R)\cos \gamma. \tag{25.2}$$

Solving for $\cos \gamma$ gives

$$\cos \gamma = \frac{R^2(1 + \cos^2 \phi_0) - L^2}{2R^2 \cos \phi_0}. \tag{25.3}$$

From the figure, it is observed that $\gamma = 180 - \delta$. We also recall, from our baseball analysis, that $\delta = \phi_0$; we emphasized this very important feature in the previous chapter. Accordingly, $\gamma = 180 - \phi_0$ and $\cos(180 - \phi_0) = -\cos \phi_0$. Substituting this result into (25.3) and carrying out a bit of algebra gives

$$R = \frac{L}{\sqrt{1 + 3\cos^2 \phi_0}}. \qquad (25.4)$$

We also know that $h/R = \sin \phi_0$. Consequently, substituting $R = h/\sin \phi_0$ into (25.4), remembering that $\sin^2 \phi_0 = 1 - \cos^2 \phi_0$, and doing a little more algebra, we obtain the formulas we want:

$$\phi_0 = \arccos \sqrt{\frac{L^2 - h^2}{L^2 + 3h^2}} \text{ and } R = \frac{1}{2}\sqrt{L^2 + 3h^2}. \qquad (25.5)$$

Therefore, with h and L known, we can calculate the values of ϕ_0 and R of the fictitious sphere.

Let us use some numbers. From the field survey we obtain $h = 15.55$ ft and $L = 68.49$ ft. So from equations (25.5) we determine that $\phi_0 = 25.0°$ and $R = 36.80$ ft. This is a great big baseball.

From table 24.2, we get $\rho = 47.57°$, $r = R \sin \rho = 27.16$ ft, $\alpha = 106.67°$, $\sigma = 84.85°$, and $S = \frac{1}{8}(10.99)R = 50.55$ ft. The distance $S = 50.55$ ft is the arc length of each of the two connection pipes. Remember that only one-fourth of the entire baseball seam is involved in our pipe connection problem. At the junction between the two pipes ($\lambda = 0°$, $\phi = 45°$), the welded flanges must be drilled to allow an angle change, $\sigma = 84.85°$, between the planes of the two circular arc pipes.

The photographs presented in figure 25.2 illustrate what we have done. The photo on the left shows half of the sphere and the $\phi_0 = 25°$ seam. The one on the right is our connection pipe with the equivalent fictitious sphere removed.

Now it is your turn to solve a problem: design and build your own pipe connection. Start with some dimensions for h and L,

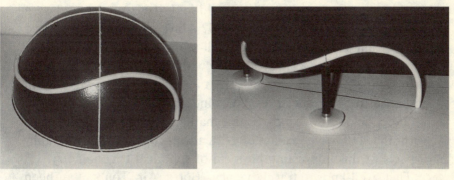

FIG. 25.2

The pipe connection problem for the case $\phi_0 = 25°$, with (left) and without (right) the equivalent sphere.

and then compute ϕ_0 and R. Forget about the sphere if you like but, by all means, visualize it.

Suggested materials of construction for your pipe connection are (a) some fairly heavy duty electrical wire from the hardware store (it is easy to bend and shape) or (b) a hula hoop you can probably get at the toy store (the material is rigid and the radius is fixed but you can easily set the correct angle change σ at the junction point and clamp it with a dowel).

If you can acquire two hula hoops, you will have enough to build the entire curve for the case $\phi_0 = 45°$. Alternatively, you may prefer to construct half of the $\phi_0 = 0°$ curve. We take a close look at this $\phi_0 = 0°$ case in the following section. For both cases, $\phi_0 = 0°$ and $\phi_0 = 45°$, the completed space curves are interesting structures. You might even classify them as attractive examples of modern art.

Halfway Around the World Along the $\phi = 0$ Curve

Here, we revisit maps and cartography. Have you ever wondered where a curve shaped like a baseball seam would go if it were projected onto a globe of the earth? Have you ever imagined how a tennis ball groove would look if it were plotted on a Mercator projection of the world? Have you ever speculated about what major city a basketball-seam-type curve would traverse if one of the seams coincided with the Greenwich meridian on a map? Well, if you are like most people you will say, in response to these questions, "Yes indeed."

We conclude our study of the mathematics of baseball seams with the following example. It features the world geography aspects of this interesting problem of spherical trigonometry.

That well-known travel agency, Bob's Air Tours, has been promoting a fantastic journey, beginning and ending in Singapore ($\lambda = 105°E$, $\phi = 0°$), that flies the $\phi_0 = 0$ route. The advertising agency has recommended to Bob's that they use something a lot more exciting than the phrase "The $\phi_0 = 0$ Route" to promote the tour.

In any event, the trip starts and stops in Singapore, which has great hotels and shopping. It is also on the equator, which makes our example easier. As you have gathered, this journey follows the $\phi_0 = 0$ curve on a sphere. As shown in figure 25.3, outbound it goes to the southwest and inbound returns from the northwest.

In the coordinate system of our example, with Singapore at $\lambda = 105°E$, the equation of the flight curve between $\lambda = 105°E$ and $\lambda = 15°E$ is given by equation (24.14) of the previous chapter. With $\phi_0 = 0$ and $q = 1$, we have

$$\phi = \arccos\left[\frac{\sin \lambda}{2 + \sin^2 \lambda}\left(1 + \sqrt{2}\, \csc \lambda\sqrt{1 + \sin^2 \lambda}\,\right)\right],$$

$$(25.6)$$

in which $\lambda = \lambda_* - 15°$, where λ_* is the longitude east of the Greenwich meridian of a point on the curve.

For the portion of the flight between $\lambda = 15°E$ and $75°W$, we employ equation (24.16). With $\phi_0 = 0$ and $\rho = \arctan\sqrt{2}$, we get

$$\phi = \arccos\left(\frac{1}{\sqrt{2}\,\cos \lambda + \sin \lambda}\right), \qquad (25.7)$$

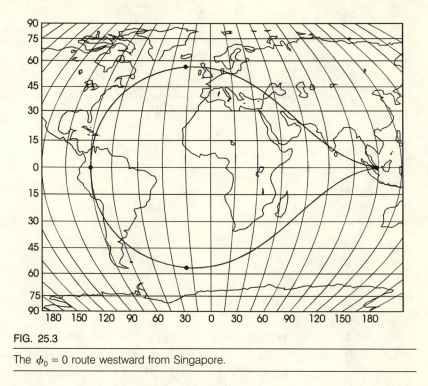

FIG. 25.3

The $\phi_0 = 0$ route westward from Singapore.

where, now, $\lambda = \lambda_* + 15°$ and λ_* is the longitude west of Greenwich of a point on the path.

From equation (24.18), with $\phi_0 = 0$, the maximum latitude occurs at $\lambda_m = \arctan(1/\sqrt{2})$; this longitude corresponds to $\lambda_* = 20.26°$W. The maximum latitude is $\phi_m = \rho = 54.74°$N and S.

The coordinates of the flight path are listed in table 25.1.

The path of this fantastic journey is shown in figure 25.3. Initially, the trip takes us across the Indian Ocean, the island of Madagascar, the tip of southern Africa, and the south Atlantic. Then we fly over Argentina, northern Chile, and Peru and reach the midpoint of the journey at a point about 400 kilometers east of Quito, Ecuador.

Homeward bound, we cross Colombia, the Caribbean, and Haiti and then follow the Gulf Stream over most of the north Atlantic. After skirting the southern tip of Newfoundland, we

TABLE 25.1

Coordinates of the path of the $\phi_0 = 0$ route
westward from Singapore

λ_*	ϕ
105°E	0.0
90°E	1.4
75°E	5.6
60°E	12.7
45°E	22.3
30°E	33.7
15 E	45.0
0	52.0
15°W	54.6
30°W	54.1
45°W	50.3
60°W	41.3
75°W	0.0

reach our maximum northern latitude approximately 800 kilometers directly south of Iceland. This is followed by the grand tour of Europe, starting with Ireland and England and ending with Greece. We then fly over northeastern Egypt, Saudi Arabia, and Yemen, and finally have another long flight over the Indian Ocean before arriving back in Singapore.

From our earlier analysis, we determined that $S/R = 13.68$ for the $\phi_0 = 0$ curve. Since our trip is only half the total and with $R = 6,370$ km, the length of our journey, S, is 43,570 kilometers. The total area enclosed by the route is one-fourth the area of the world, or 127.5 million square kilometers.

The map showing our journey in figure 25.3 is the Robinson projection of the world. If you want to know more about this projection—and a great many more—you should consult the excellent album of map projections by Snyder and Voxland (1989).

Before we leave the subject, a word of caution. Check out Bob's Air Tours before you go on this long and expensive trip. For example, they advertise "We Take You All Around Africa?" Do they ever! Look at the map.

Finally, why not take an hour or two sometime and make a plot of Bob's "$\phi_0 = 0$ Route" *eastward* from Singapore. Where does it go? Where in the world are the points of maximum latitude? This assignment is easy; you have all the numbers in table 25.1.

This is enough! We have carried out extensive analyses of topics relating to baseball seams and tennis ball grooves. Topics involving basketball seams do not present much of a problem. We shall save the investigation of golf ball dimples for another day.

26

Lengths, Areas, and Volumes of All Kinds of Shapes

An intriguing book you might want to examine sometime is *Engineering in History*, by Kirby et al. (1990). One thing becomes quite apparent as you read this informative text: for thousands of years, humans have been making calculations about things they build.

The Great Pyramid of Cheops in Egypt is an excellent example. Constructed over 4,500 years ago, this structure had a height H of 480 feet and an average side length L of 750 feet. So its volume $V = HL^2/3$ was approximately 90 million cubic feet. This required 2.1 million blocks of limestone, each block measuring 3.5 feet on a side and weighing over 6,000 pounds.

Of course, the ancient Egyptians did not use our units of pounds and feet; they had their own units of measurements. Never mind. The point is that in order to build the enormous pyramids, they had to compute numerous quantities. With these observations, we begin our own calculations about some interesting shapes.

The Washington Monument

Although it may not be the most beautiful structure in the nation's capital, the Washington Monument is certainly the most familiar, the tallest, and probably the most impressive. It is

555 feet 5 inches in height and is made entirely of marble and granite. As its name implies, this magnificent structure is a memorial to America's first president.

Following many years of deliberation and planning, the cornerstone of the monument was laid on July 4, 1848 during the term of office of President James Polk. The original design called for a very elaborate Parthenon-type structure as the base for a 500-foot square shaft with a star on the top. The plan called for the shaft to be built first; the base was to be constructed later on.

Things went along fairly well for the next five years. Unfortunately, after the structure had risen to a height of 152 feet, funds ran out in 1854 and for the ensuing twenty-five years there was no further progress on construction. Eventually, in 1880, Congress appropriated the necessary funds and work on the monument was resumed. The earlier plan for the fancy base was discarded and the design of the shaft was somewhat modified. The monument was finally completed in December 1884. A photograph of the imposing structure is shown in Figure 26.1.

At the top of the 500-foot tapered column is a 55-foot pyramidion and on top of that is a nine-inch aluminum capstone. An elevator operates to the 500-foot level but there are 897 steps for people who prefer to walk up and down.

What Is the Surface Area of the Washington Monument?

The architectural shape of the monument is an obelisk; notable examples of such structures are seen among the monuments of ancient Egypt. A drawing of the structure is shown in figure 26.2, in which the proportions are substantially distorted so that dimensions can be displayed more clearly.

We now carry out calculations to determine the surface area.

The Sides of the Main Column. There is a slight taper to the main column. From the dimensions given in figure 26.2, we easily determine that the slant length of the sides is $\sqrt{(500.4)^2 + (10.3)^2} = 500.5$ ft. So the area of each of the four

FIG. 26.1

The Washington Monument. (Photograph provided by National Park Service.)

trapezoidal sides is given by $(1/2)(55.0 + 34.4)(500.5) = 22,372$ ft^2. Consequently, the total surface area of the main column is $A_1 = 89,488$ ft^2.

The Pyramidion at the Top. The slant length of the sides is $\sqrt{(55.0)^2 + (17.2)^2} = 57.6$ ft. Therefore the area of each of the four triangular sides is $(1/2)(34.4)(57.6) = 991$ ft^2. Hence, the total area of the pyramidion is $A_2 = 3,964$ ft^2.

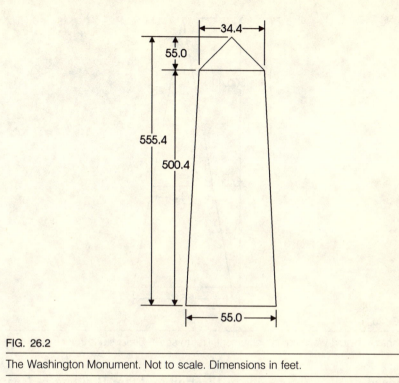

FIG. 26.2

The Washington Monument. Not to scale. Dimensions in feet.

Total Surface Area. If we now add A_1 and A_2, we obtain $A = 93,452$ ft². This is the total surface area of the monument.

What Is the Total Volume of the Washington Monument?

To simplify our computations for the volume, we (a) temporarily remove the pyramidion at the top and (b) extend the tapered sides of the main column until they meet high in the sky. The geometry of this situation is shown in figure 26.3. Next, utilizing the geometrical property of similar triangles, we get the equation

$$\frac{H - 500.4}{34.4} = \frac{H}{55.0}.$$

We easily solve this equation to obtain $H = 1,336$ ft. This height is more than double the actual height of the monument, and

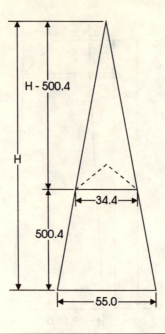

FIG. 26.3

A sharp-pointed Washington Monument. Not to scale. Dimensions in feet.

indeed exceeds the height of the Empire State Building in New York.

Back to our computations: The volume of a pyramid is given by the expression $V = (1/3)BH$, in which B is the area of the base and H is the height.

The Main Column. First, we calculate the volume of the entire—but mythical—1,336-foot pyramid. Accordingly,

$$V_1 = (1/3)(55.0)^2(1,336) = 1,347,133 \text{ ft}^3.$$

Next, we compute the volume of the top—but also mythical—pyramid of height $1,336 - 500.4 = 835.6$ ft. We have

$$V_2 = (1/3)(34.4)^2(835.6) = 329,605 \text{ ft}^3.$$

So the volume of the main column is $V_1 - V_2 = 1,017,528 \text{ ft}^3.$

The Pyramidion at the Top. The volume of the 55-foot pyramid at the top of the monument is

$$V_3 = (1/3)(34.4)^2(55.0) = 21,695 \text{ ft}^3.$$

Total Volume. From the preceding results, we easily obtain the total volume of the monument: $V = 1,039,223 \text{ ft}^3$.

How Many Golf Balls Would Fit Inside the Washington Monument?

This is something you have always wanted to know, right? We hasten to add that we are not being frivolous or disrespectful when we ask this question. As we shall see in a moment, we are simply utilizing this physical setting to pose an interesting problem in solid geometry.

It is assumed that the entire structure is empty (i.e., no stairs, no elevator) and that the walls are of zero thickness (in fact, they are 15 feet thick at the base). So the total volume is $V = 1,039,223 \text{ ft}^3$. The diameter of a golf ball is $D = 1.65$ in and so its radius is $R = 0.825$ in. Consequently, the volume of a golf ball is $v = (4/3)\pi R^3 = 2.352 \text{ in}^3 = 0.001361 \text{ ft}^3$.

Here comes the problem. No matter how you arrange a collection of uniform-diameter spheres in a given space, only a certain fraction of the space will be occupied by the spheres (i.e., golf balls); the remaining fraction—the so-called pore or void space—will be occupied by air.

So we have the following problem in solid geometry. What arrangement of uniform-diameter spheres, in contact with other spheres, will provide (a) the *least dense* and (b) the *most dense* packing?

Well, it turns out that the least dense is a *cubic* packing. It has a porosity (ratio of pore space to total space) of $n = 0.4764$. On the other hand, the most dense is termed the *rhombohedral* packing with porosity $n = 0.2595$.

This kind of information concerning the porosity of what are called "packed beds" is very important to chemical and petroleum engineers in the design of columns and tanks for the processing of petrochemical products. For these purposes, engineers usually select porosity values in the range $n = 0.35$ to 0.38.

Back to our problem: Using a total volume $V = 1,039,223$ ft^3 and selecting a porosity $n = 0.37$ means that the volume occupied by the pores (i.e., air) is $V_p = (0.37)(1,039,223) = 384,513$ ft^3. Consequently, the volume occupied by the golf balls (i.e., spheres) is $V_s = 654,710$ ft^3.

Since the volume of a single golf ball is $v = 0.001361$ ft^3, the total number of balls that would fit inside the Washington Monument is

$$N = \frac{654.710 \times 10^3}{1.361 \times 10^{-3}} = 480 \times 10^6 = 480 \text{ million.}$$

This is quite a few golf balls.

A RESEARCH ASSIGNMENT.   Perhaps you would like to examine more closely the solid geometry problem concerning the least dense and the most dense packing arrangements of uniform-diameter spheres. Here's a suggestion for some simple experiments: glue or tape some ping-pong balls together to construct a few layers of spheres in various patterns. Then see how the layers fit together to provide loose and dense packings.

Can you express these geometrical patterns in terms of the mathematics needed to compute the porosity? Suggested references are Steinhaus (1969) and Greenkorn (1983).

Blimps

We are now going to calculate the volumes and surface areas of a few things with which we are all familiar. We start with blimps. These are the big airships we frequently see sailing over football stadiums, along crowded beaches and numerous other places.

The largest airship—blimp—ever constructed was the U.S. Navy's ZPG 3-W. Four of these were built for the navy by Goodyear in the late 1950s. We shall calculate some things about

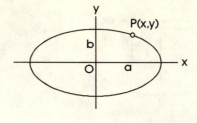

FIG. 26.4

Definition sketch for an ellipse. We assume that a blimp has an elliptical shape.

these enormous blimps after we develop the mathematical relationships needed for our calculations.

Ellipses, Ellipsoids, and Blimps

The main tools we require for our analysis are obtained from analytic geometry and elementary calculus. However, if you have not studied these subjects or if you have forgotten them, do not be discouraged. The equations and formulas we derive are easy to understand and apply.

An ellipse with semimajor axis of length a and semiminor axis of length b is shown in figure 26.4. From analytic geometry we know that the equation of an ellipse is

$$\frac{x^2}{a^2} + \frac{y^2}{b^2} = 1, \tag{26.1}$$

where, as shown in the figure, x and y are the coordinates of any point P on the curve.

Now suppose we rotate the ellipse about its major axis (i.e., the x-axis). We create a body of revolution called a *prolate ellipsoid* (or prolate spheroid). It looks like a watermelon. Now the general equation of an ellipsoid is

$$\frac{x^2}{a^2} + \frac{y^2}{b^2} + \frac{z^2}{c^2} = 1, \tag{26.2}$$

in which the z-axis passes through the origin perpendicular to the x-y plane; c is the half-length in the z-direction. In the case of

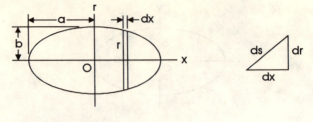

FIG. 26.5

Definition sketch for a prolate ellipsoid.

the *prolate* ellipsoid, $c = b$. If we let $r^2 = y^2 + z^2$, then (26.2) becomes

$$\frac{x^2}{a^2} + \frac{r^2}{b^2} = 1, \tag{26.3}$$

which is the equation of the prolate ellipsoid.

Alternatively, suppose we rotate the ellipse about its minor axis (i.e., the y-axis). This time we generate a so-called *oblate ellipsoid* (or oblate spheroid). This has somewhat the shape of a pumpkin. Indeed, the earth itself is an oblate ellipsoid because the length of its polar diameter, $2b$ is slightly less than the length of its equatorial diameter, $2a$. This time, $c = a$. If we let $r^2 = x^2 + z^2$, then (26.2) becomes

$$\frac{r^2}{a^2} + \frac{y^2}{b^2} = 1. \tag{26.4}$$

This is the equation of the oblate ellipsoid. In this chapter, we are interested in only the prolate ellipsoid, so we set equation (26.4) aside.

How to Calculate the Volume of a Blimp

First we calculate the volume of a prolate ellipsoid by using integral calculus. With reference to figure 26.5, the volume of the thin disk is $dV = \pi r^2 dx$. Clearly, the summation of the volumes of all the thin disks, from $x = -a$ to $x = +a$, gives the volume of the entire ellipsoid. Now, from (26.3) we have $r^2 = b^2(1 - $

x^2/a^2). So we set up the following problem in integral calculus and obtain a relatively easy solution:

$$V = \pi b^2 \int_{-a}^{+a} (1 - x^2/a^2)\, dx = (4\pi/3)ab^2. \tag{26.5}$$

We assume that the blimp has the shape of a prolate ellipsoid. Then, if $L = 2a$ is the length of the blimp and $D = 2b$ is its diameter, equation (26.5) becomes simply

$$V = \frac{\pi}{6} LD^2. \tag{26.6}$$

With this formula we easily compute the volume of our blimp.

How to Calculate the Surface Area of a Blimp

This problem is a bit more complicated. Again, with reference to figure 26.5, the surface area of the narrow strip along the outer edge of the thin disk is $dA = 2\pi r\, ds$. It should be clear that $ds = \sqrt{(dx)^2 + (dr)^2}$. Accordingly, the summation of the areas of all the narrow strips, from $x = -a$ to $x = +a$, gives the total surface area of the ellipsoid. So we have

$$A = 2\pi \int_{-a}^{+a} r\, ds = 2\pi \int_{-a}^{+a} r \sqrt{1 + \left(\frac{dr}{dx}\right)^2}\, dx. \tag{26.7}$$

From equation (26.3) we obtain expressions for r and for dr/dx. Substituting these into (26.7) and carrying out the integration yields the result

$$A = 2\pi ab\left[\lambda + \frac{1}{\sqrt{1 - \lambda^2}} \arcsin\sqrt{1 - \lambda^2}\right]; \quad \lambda = \frac{b}{a}. \tag{26.8}$$

Again using $L = 2a$ and $D = 2b$ as the blimp's length and diameter, respectively, we obtain the final answer:

$$A = \frac{\pi LD}{2} \left[\lambda + \frac{1}{\sqrt{1 - \lambda^2}} \arcsin\sqrt{1 - \lambda^2} \right]; \; \lambda = \frac{D}{L}.$$

$$(26.9)$$

This formula gives us the surface area of the blimp.

The Navy's ZPG 3-W Marine Patrol Airship

This was the largest airship ever constructed. Its length was $L = 403$ ft and its diameter $D = 85$ ft. It is reported that its volume was about 1.5 million cubic feet.

If we substitute these numerical values of L and D into equation (26.6), we obtain the computed volume $V = 1,525,000$ ft^3, which is pretty close to the reported value. Substituting the numbers into (26.9) gives the calculated surface area $A = 86,200$ ft^2. This is an area of almost exactly 2.0 acres.

Lifting Capacity of Blimps

The ability of a blimp to lift not only its own weight but also the weight of an appreciable payload is due to the fact that a large fraction of its total volume contains a gas that is lighter than air. In the old-time dirigibles, especially those of the Germans, hydrogen was used as the buoyant gas. Hydrogen, of course, is by far the lightest known gas but it is also extremely dangerous to use. The explosion of its hydrogen tanks is what caused the famous *Hindenburg* to crash and burn at Lakehurst, New Jersey in May 1937.

Because hydrogen is so hazardous, many of the early dirigibles and all of the present-day blimps use helium as the buoyant gas.

QUESTION FOR YOUR COFFEE BREAK. If the specific weight of air is $w_1 = 0.077$ lb/ft^3 and the specific weight of helium is $w_2 =$

0.011 lb / ft³, and if helium occupied the entire volume of the ZPG 3-W blimp, what was the blimp's total lifting capacity?

Answer. 100,650 pounds.

Footballs

There are rigid specifications concerning the proper size and shape of footballs. For example, the National Football League stipulates that the football be

11 to 11.25 inches in length

21.5 inches circumference (round part)

28.5 inches circumference (long part)

"... it shall have the shape of a prolate spheroid"

From these specifications, it is easy to determine the following dimensions of a football: $a = 5.50$ in, $b = 3.40$ in, $L = 2a = 11.0$ in, and $D = 2b = 6.8$ in. If we substitute these numbers into equations (26.6) and (26.9), we obtain a football's volume V and surface area A, respectively. The answers are $V = 266$ in³ and $A = 208$ in².

PROBLEM   In spite of the NFL's stipulation that the football have the shape of a prolate spheroid, it simply doesn't look like one. It does not resemble a small blimp; its ends are much too pointed.

So, using one or the other of the following profile equations of a football, set up the integrals that lead to formulas for the volume and surface area of a football:

Arc of a circle

$$r = \sqrt{R^2 - x^2} - (R - b), \tag{26.10}$$

where $R = 6.15$ in, $b = 3.40$ in, $R - b = 2.75$ in.

Parabola

$$r = b\left(1 - \frac{x^2}{a^2}\right), \tag{26.11}$$

where $a = 5.50$ in, $b = 3.40$ in.

In your analysis, equation (26.10) or (26.11) would replace (26.3), which, of course, is the equation of an ellipse. Both (26.10) and (26.11) provide nice pointed ends for the football.

Doughnuts

Toward the end of the third century, a mathematician named Pappus of Alexandria discovered two rules or relationships that are very useful for the computation of the volumes and areas of solids. Unfortunately, these discoveries of Pappus were lost for over 1,200 years. Then, in the seventeenth century, the famous German mathematician Johannes Kepler (1591–1630) rediscovered them and put them to use.

These two relationships or rules of Pappus are

> *Volume rule*: The volume of any solid generated by the revolution of a plane area about an external axis in its plane is equal to the product of the area of the generating figure and the distance its center of gravity moves

> *Surface area rule*: The area of any surface generated by the revolution of a plane curve about an external axis in its plane is equal to the product of the generating curve and the distance its center of gravity moves

An Application of the Relationships of Pappus

In figure 26.6, a torus—a solid shaped like a doughnut or a tire—is shown with primary radius R and secondary radius a. What is its volume V and surface area A?

Employing the volume rule, we obtain

$$V = \pi a^2 \times 2\pi R = 2\pi^2 a^2 R, \qquad (26.12)$$

and using the surface area rule, we get

$$A = 2\pi a \times 2\pi R = 4\pi^2 a R. \qquad (26.13)$$

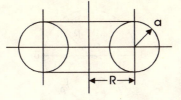

FIG. 26.6

Definition sketch for a torus.

A NUMERICAL EXAMPLE. The largest tires ever made were manufactured by Goodyear for enormous earth-moving trucks. These tires had an outer diameter of 12 feet and weighed 12,500 pounds. Assuming an inner diameter of 6 feet, it is easily established that $R = 4.5$ ft and $a = 1.5$ ft.

From equations (26.12) and (26.13), we determine that the volume of this gigantic tire was $V = 200$ ft^3 and the area was 266 ft^2.

A Goat Tied to a Fence: The Involute of a Circle

Here is an interesting little problem to think about. A goat is tied to one end of a rope of length R. The other end of the rope is tied to a post on the outside of a circular fence of radius a. In terms of R and a, what is the total area over which the goat can graze the grass outside the fence?

The problem should be clear enough. As shown in figure 26.7, in zone I the full length of the rope, R, determines the goat's grazing region. This provides the semicircular area $A_1 = \pi R^2/2$.

However, in zones II the effective length of the rope is reduced because a portion of it may be wrapped along the circular fence. Indeed, R, the length of the rope, is the arc length of the fence measured from post P to either of points A. The shortening of the rope results in a reduction of the grazing area. So the problem is to determine the areas of zones II. To calculate these areas, we need to know the equation of the curve AB.

FIG. 26.7

A goat grazing on the grass outside a circular fence. Curve *AB* is called the involute of a circle.

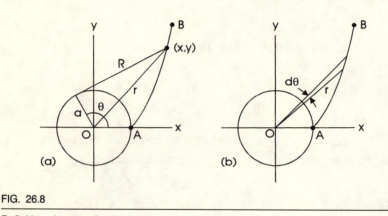

FIG. 26.8

Definition sketches for the involute of a circle.

A definition sketch of our problem is displayed in figure 26.8(a). It is not difficult to show that the so-called parametric equations of the curve *AB* are the following:

$$x = a(\cos \theta + \theta \sin \theta), \tag{26.14}$$

$$y = a(\sin \theta - \theta \cos \theta). \tag{26.15}$$

These equations define a curve called the *involute of a circle*. In this regard you might want to refer to books by Lawrence (1972) and by Gellert et al. (1977).

Use of the trigonometric identity $\sin^2 \theta + \cos^2 \theta = 1$ and a bit of algebra reduces equations (26.14) and (26.15) to the simple form

$$r = a\sqrt{1 + \theta^2}. \tag{26.16}$$

This is the equation of curve AB in terms of polar coordinates.

With reference to figure 26.8(b), we note that the area of the small triangle is $dA = (1/2)r(rd\theta)$. Accordingly, the area swept out by the radius r as it rotates from $\theta = 0$ to $\theta = \theta$ is

$$A' = \frac{1}{2}\int_0^\theta r^2\, d\theta. \tag{26.17}$$

Substituting (26.16) into (26.17) and integrating gives

$$A'' = \frac{a^2}{2}\left(\theta + \frac{1}{3}\theta^3\right). \tag{26.18}$$

However, the area given by this equation includes a circular area *inside* the fence. We need to subtract this area. Then we utilize the relationship $R = a\theta_0$, where θ_0 is the value of θ corresponding to point B. This gives the area of one of the zones II. Since there are two such zones, we finally get the area of zones II, $A_2 = R^3/3a$. Adding this to the area of zone I provides the final answer:

$$A = R^2\left(\frac{\pi}{2} + \frac{R}{3a}\right). \tag{26.19}$$

This is the total area over which the goat can graze grass.

Suppose, for example, that the length of the rope is $R = \pi a$ (i.e., the rope winds entirely around to the opposite side of the fence). Then $A = (5/6)\pi R^2$. This is a rather interesting result.

More Information about the Involute of a Circle

By definition, the involute of a circle is the curve traced by the end of a string as it is tautly unwound from a circular spool of radius a. The equation of the curve, expressed in polar coordinates, is given by (26.16).

It is easy to determine the length of this curve. Over a very short distance, the length is given by $ds = \sqrt{(dx)^2 + (dy)^2}$. Expressing this in terms of the parameter θ, we have

$$ds = \sqrt{\left(\frac{dx}{d\theta}\right)^2 + \left(\frac{dy}{d\theta}\right)^2} \, d\theta. \tag{26.20}$$

Substituting the derivatives of equations (26.14) and (26.15) into this expression yields $ds = a\theta d\theta$. Integrating this relationship from $\theta = 0$ to $\theta = \theta$ yields the very simple answer,

$$S = \frac{1}{2}a\theta^2, \tag{26.21}$$

where S is the length of the curve.

EXAMPLE 1. In our goat and fence problem, suppose that $a = 10$ ft and that $R = \pi a$ (i.e., the rope extends halfway around the fence). So $\theta = 180° = \pi$ radians. Then, with reference to figure 26.7 and using (26.21), we determine that the length of the curve from A to B is $S = 49.3$ ft.

EXAMPLE 2. Suppose a thread is tautly unwound from a spool with radius $r = 1.0$ in. What is the length of the curve traced by the end after two turns of thread from the spool? Since one turn of the thread corresponds to $360° = 2\pi$ radians, then $\theta = 2(2\pi) = 4\pi$. Substituting this into (26.21), with $a = 1.0$ in, gives $S = 79.0$ in. A graphical display of this example is presented in figure 26.9.

The general definition of an involute is that it is the curve traced by the end of a thread as it is tautly unwound from a given curve.

Some examples: (a) the involute of a catenary is a tractrix, (b) the involute of a parabola is a semicubic parabola, (c) the involute of a logarithmic spiral is an identical logarithmic spiral, (d) the involute of a cycloid is an identical cycloid, and so on.

Along these lines, Hoffman (1998) presents an interesting analysis of a generalization of the goat and circular fence prob-

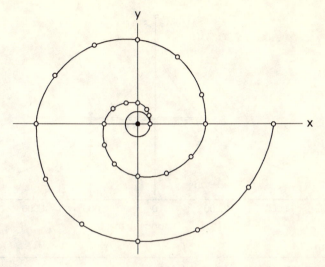

FIG. 26.9

The involute of a circle. $a = 1.0$ in. Small circles identify 30° intervals. Length of curve $S = 79.0$ in.

lem. He provides the solution corresponding to a fence defined by any smooth convex curve. As an example, he computes the grazing area for the case in which the fence has the shape of a catenary.

Mayon Volcano and the Hershey Chocolate Kiss

We have two more problems; they both involve integral calculus. The first is to compute the volume of the Mayon Volcano in the Philippines and the second is to calculate the volume of a Hershey chocolate kiss. Mathematically speaking, the two problems are essentially the same.

In chapter 20, we consider the geometry of the Mayon Volcano. This beautiful active volcano has a height of 2,460 meters and covers an area of about 300 square kilometers near sea level in the southern part of Luzon in the Philippines.

This mountain is so beautifully symmetric that, in the past, numerous mathematical equations have been proposed to de-

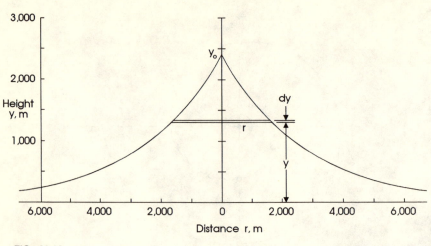

FIG. 26.10

Mathematical profile of the Mayon Volcano.

scribe its shape. One such equation is

$$r = c \sinh\left(\frac{y_0 - y}{c}\right), \tag{26.22}$$

where, as shown in figure 26.10, r is the horizontal distance from the vertical y-axis, $y_0 = 2{,}460$ m is the elevation of the peak, y is the elevation at horizontal distance r, and $c = 800$ m is a constant. A plot of equation (26.22) is shown in figure 26.10.

Here is our question: assuming that equation (26.22) describes the profile of the Mayon Volcano, what is the volume of the mountain?

Once again we have a problem in integral calculus. The volume of the thin disk shown in figure 26.10 is $dV = \pi r^2 dy$. If we add up the volumes of all the thin disks from $y = 0$ to $y = y_0$, we have the volume of our mountain. That is,

$$V = \pi \int_0^{y_0} r^2 \, dy. \tag{26.23}$$

Substituting (26.22) into (26.23) gives

$$V = \pi c^2 \int_0^{y_0} \sinh^2\left(\frac{y_0 - y}{c}\right) dy. \tag{26.24}$$

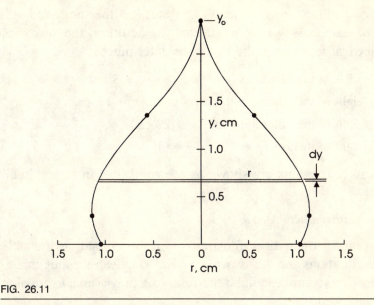

FIG. 26.11

Mathematical profile of a chocolate kiss.

At this point, we need a table of integrals to evaluate the right-hand side. Without much difficulty we obtain the following result:

$$V = \frac{\pi c^3}{2}\left(\sinh\frac{y_0}{c}\cosh\frac{y_0}{c} - \frac{y_0}{c}\right).$$ (26.25)

Substituting $y_0 = 2{,}460$ m $= 2.460$ km and $c = 800$ m $= 0.80$ km into this equation provides the desired answer: $V = 91.7$ km^3. This is about 22.0 cubic miles.

Hershey Chocolate Kiss

A similar analysis is made to determine the volume of a chocolate kiss. The profile of the kiss is shown in figure 26.11.

Results of Laboratory Work

Some preliminary laboratory work was carried out involving a sack of Hershey's milk chocolate kisses containing 35 kisses. Utilizing a measuring cup, it was determined that the 35 kisses displaced 140 cubic centimeters of water. So the measured volume of each kiss was $V = 4.0$ cm^3. In addition, the following geometrical features of the kiss were determined:

1. Radius of the kiss at its base ($y = 0$, $r = 1.05$ cm)
2. Height of the maximum thickness point ($y = 0.3$ cm)
3. Height of the inflection point ($y = 1.3$ cm)
4. Height of the kiss ($y = 2.2$ cm, $r = 0$)

These four points are shown as the solid dots in figure 26.11.

The Mathematical Model

Several mathematical functions were examined as possible shape equations for the chocolate kiss. They were either inadequate or too complicated. So the following polynomial expression was employed:

$$r = k_0 + k_1 y + k_2 y^2 + k_3 y^3. \qquad (26.26)$$

This equation contains four constants which are easily determined from the four measurements indicated above.

For example, the first derivative of equation (26.26) is $dr/dy = k_1 + 2k_2 y + 3k_3 y^2$. Since this quantity represents the slope of the profile, we know that $dr/dy = 0$ at the maximum thickness point. Likewise, the second derivative is $d^2r/dy^2 = 2k_2 + 6k_3 y$. This quantity represents the curvature of the profile and this must be zero at the inflection point. In this way, the numerical values of the four constants were computed. The results are $k_0 = 1.05$, $k_1 = 0.592$, $k_2 = -1.115$, and $k_3 = 0.286$.

We construct our integral calculus problem in the same way as in our analysis of Mount Mayon. The volume of the thin disk shown in figure 26.11 is $dV = \pi r^2 dy$. Therefore, the total volume is

$$V = \pi \int_0^{y_0} r^2 \, dy. \qquad (26.27)$$

We now substitute equation (26.26) into (26.27) and integrate term by term. The algebra is somewhat complicated, but there is nothing difficult. For such integration it is helpful to know that

$$\int x^n \, dx = \frac{x^{n+1}}{n+1} + C, \; n \neq -1, \tag{26.28}$$

where C is a constant of integration. This is the general rule for determining the integral of a variable x when raised to a power n. This rule fails when $n = -1$. In this case, we have

$$\int \frac{dx}{x} = \log_e x + C. \tag{26.29}$$

The final algebraic answer provided by integrating (26.27) is a lengthy expression containing ascending powers of y_0, the height of the kiss, and various combinations of the four numerical constants.

The numerical result of our mathematical model and computation is $V = 4.15$ cm^3. This is remarkably close to the measured value of the volume of a Hershey's chocolate kiss, $V = 4.0$ cm^3.

References

Abramowitz, M., and I. A. Stegun (1965). *Handbook of Mathematical Functions*. New York: Dover Publications.

Addison, P. S. (1997). *Fractals and Chaos: An Illustrated Course*. Philadelphia: Institute of Physics Publishing.

Aravind, P. K. (1994). "A symmetrical pursuit problem on the sphere and the hyperbolic plane." *The Mathematical Gazette* 78:30–36.

Audouze, A., and G. Israël, eds. (1985). *The Cambridge Atlas of Astronomy*. Cambridge, England: Cambridge University Press.

Austin, A. L., and J. W. Brewer (1971). "World population growth and related technical problems." *Technological Forecasting and Social Change* 3:23–49.

Austin, J. D., and F. B. Dunning (1988). "Mathematics of the rainbow." *Mathematics Teacher* 81:484–488.

Bagrow, L. (1985). *History of Cartography*. 2nd ed. Chicago: Precedent Publishing.

Bailey, D. H., J. M. Borwein, P. B. Borwein, and S. Plouffe (1997). "The quest for pi." *The Mathematical Intelligencer* 19:50–56.

Banks, R. B. (1994). *Growth and Diffusion Phenomena: Mathematical Frameworks and Applications*. Heidelberg: Springer-Verlag.

Barenblatt, G. I. (1987). *Dimensional Analysis*. New York: Gordon and Breach.

Becker, G. F. (1905). "A feature of Mayon Volcano." *Proceedings of the Washington Academy of Science* 7:277–282.

Beckmann, P. (1977). *A History of Pi*. Boulder, Colo.: Golem Press.

Beiler, A. H. (1964) *Recreations in the Theory of Numbers*. New York: Dover Publications.

Bell, E. T. (1956). "The queen of mathematics." In *The World of Mathematics I*, ed. J. R. Newman, 498–518. New York: Simon and Schuster.

Bentley, W. A., and W. J. Humphreys (1962). *Snow Crystals*. New York: Dover Publications.

Berggren, L., P. Borwein, and J. Borwein (1997). *Pi: A Source Book*. New York: Springer-Verlag.

Boas, M. L. (1983). *Mathematical Methods in the Physical Sciences*. 2nd ed. New York: John Wiley.

Boyce, W. E., and R. C. DiPrima (1996). *Elementary Differential Equations and Boundary Value Problems*. 6th ed. New York: John Wiley.

Boyer, C. B. (1987). *The Rainbow: From Myth to Mathematics*. Princeton, N.J.: Princeton University Press.

———— (1991). *A History of Mathematics*. 2nd ed. Revised by U. C. Merzbach. New York: John Wiley.

Brown, L. A. (1949). *The Story of Maps*. Boston: Little, Brown and Co.

Bruce, W. J. (1978). "Crazy roller coasters." *Mathematics Teacher* 71:45–49.

Byalko, A. V. (1987). *Our Planet—The Earth*. Moscow: Mir Publishers.

Çambel, A. B. (1993). *Applied Chaos Theory: A Paradigm for Complexity*. Boston: Academic Press.

Campbell, D. M., and J. C. Higgins (1984). *Mathematics: People, Problems, Results*. Belmont, Calif.: Wadsworth International.

Chorley, R. J., ed. (1972). *Spatial Analysis in Geomorphology*. London: Methuen and Co.

Chorley, R. J., and L. S. D. Morley (1959). "A simplified approximation for the hypsometric integral." *Journal of Geology* 67:566–571.

Coale, A. J. (1974). "The history of human populations." *Scientific American* 231:50–51.

Coffey, W. J. (1981). *Geography: Towards a General Spatial Systems Approach*. New York: Methuen and Co.

Cogley, J. G. (1985). "Hypsometry of the continents." *Annals of Geomorphology* 53:1–48.

Coxeter, H. S. M. (1996). "The trigonometry of Escher's woodcut *Circle Limit III*." *Mathematical Intelligencer* 18(4):42–46.

Coxeter, H. S. M., M. Emmer, R. Penrose, and M. L. Teuber (1986). *M. C. Escher: Art and Science*. New York: Elsevier Science Publishing Co.

Crampton, W. (1990). *Flags of the World. A Pictorial History*. New York: Dorset Press.

Davis, H. T. (1962). *Introduction to Nonlinear Differential and Integral Equations*. New York: Dover Publications.

Davis, P. J. (1961). *The Love of Large Numbers*. New York: Random House.

Davis, P. J., and R. Hersh (1981). *The Mathematical Experience*. Cambridge, Mass.: Birkhäuser.

de Blij, H. J., and P. O. Muller (1996). *Physical Geography of the Global Environment*. 2nd ed. New York: John Wiley.

Deevey, E. S. (1960). "The human population." *Scientific American* 203:194–205.

Denton, G. H., and T. J. Hughes, eds. (1981). *The Last Great Ice Sheets*. New York: John Wiley.

Diem, K., and G. Lentner, eds. (1970). *Scientific Tables*. 7th ed. Basle, Switzerland: Ciba-Geigy Ltd.

Dingle, A. N., and Y. Lee (1972). "Terminal fallspeeds of raindrops." *Journal of Applied Meteorology* 11:877–879.

Dörrie, H. (1965). *100 Great Problems of Elementary Mathematics*. New York: Dover Publications.

Dunham, W. (1990). *Journey Through Genius: The Great Theorems of Mathematics*. New York: John Wiley.

Dunlap, R. A. (1997). *The Golden Ratio and Fibonacci Numbers*. River Edge, N.J.: World Scientific Publishing.

Dunn, P., ed. (1980). *Mathematical Bafflers*. New York: Dover Publications.

Edwards, D., and M. Hamson (1990). *Guide to Mathematical Modelling*. Boca Raton, Fla.: CRC Press.

Edwards, I. E. S. (1993). *The Pyramids of Egypt*. Rev. ed. New York: Penguin Books.

Feller, W. (1940). "On the logistic law of growth and its experimental verifications in biology." Chapter 4 in *Applicable Mathematics of Nonphysical Phenomena*, eds. F. Olivera-Pinto and B. W. Connolly, 123–138. Chichester, U.K.: Ellis Horwood.

Fernie, J. D. (1991, 1992). "The shape of the earth." *American Scientist* 79:108–110, 393–395; 80:125–127.

Forray, M. J. (1986). *Variational Calculus in Science and Engineering*. New York: McGraw-Hill.

Furlong, W., and B. McCandless (1981). *So Proudly We Hail. The History of the United States Flag*. Washington, D.C.: Smithsonian Institution Press.

Gardner, M. (1992). *Fractal Music, Hypercards and More*. New York: W. H. Freeman.

Gellert, W., M. Küstner, M. Hellwich, and H. Kästner (1977). *The VNR Concise Encyclopedia of Mathematics*. New York: Van Nostrand Reinhold.

Ghyka, M. (1978). *The Geometry of Art and Life*. New York: Dover Publications.

Glaister, P. (1996). "Golden earrings." *Mathematical Gazette* 80:224–225.

Goldberg, S. (1983). *Probability in Social Science*. Cambridge, Mass.: Birkhäuser.

Greenkorn, R. A. (1983). *Flow Phenomena in Porous Media*. New York: Marcel Dekker.

Greenler, R. (1980). *Rainbows, Halos and Glories*. Cambridge, U.K.: Cambridge University Press.

Griliches, Z. (1960). "Hybrid corn and the economics of innovation." *Science* 132:275–280.

Gross, M. G. (1985). *Oceanography*. 5th ed. Columbus, Ohio: Charles E. Merrill.

Harte, J. (1985). *Consider a Spherical Cow*. Los Altos, Calif.: William Kaufmann.

Hayes, B. (1984). "On the ups and downs of hailstone numbers." *Scientific American* 250:13–17.

Hoffman, M. E. (1998). "The bull and the silo: an application of curvature." *American Mathematical Monthly* 105:55–58.

Huntley, H. E. (1970). *The Divine Proportion: A Study in Mathematical Beauty*. New York: Dover Publications.

Isenberg, C. (1992). *The Science of Soap Films and Soap Bubbles*. New York: Dover Publications.

Jolley, L. B. W. (1961). *Summation of Series*. New York: Dover Publications.

Kappraff, J. (1990). *Connections: The Geometric Bridge Between Art and Science*. New York: McGraw-Hill.

Kepler, J. ([1611] 1966). *The Six-Cornered Snowflake*. Ed. Colin Hardie. Reprint, Oxford, U.K.: Clarendon Press.

Keyfitz, N. (1966). "How many people have lived on earth?" *Demography* 3:581–582.

——— (1968). *Introduction to the Mathematics of Population.* Reading, Mass.: Addison-Wesley.

Kirby, R. S., S. Withington, A. B. Darling, and F. G. Kilgour (1990). *Engineering in History.* New York: Dover Publications.

Knight, C., and N. Knight (1973). "Snow crystals." *Scientific American* 228:100–107.

Kosko, E. (1987). "Three-dimensional pipe joining." In *Mathematical Modelling: Classroom Notes in Applied Mathematics,* ed. M. S. Klamkin. Philadelphia: SIAM.

Lagarias, J. C. (1985). "The $3x + 1$ problem and its generalizations." *American Mathematical Monthly* 92:3–23.

Lamb, H. (1945). *Hydrodynamics.* 6th ed. New York: Dover Publications.

Lawrence, J. D. (1972). *A Catalog of Special Plane Curves.* New York: Dover Publications.

Lines, M. E. (1986). *A Number for Your Thoughts.* Bristol, U.K.: Adam Hilger.

——— (1990). *Think of a Number.* Bristol, U.K.: Adam Hilger.

Lockwood, E. H. (1961). *A Book of Curves.* Cambridge, U.K.: Cambridge University Press.

Lord, N. (1995). "Balancing and golden rectangles." *The Mathematical Gazette* 79:573–574.

MacDonald, N. (1989). *Biological Delay Systems: Linear Stability Theory.* Cambridge, U.K.: Cambridge University Press.

Mandelbrot, M. (1982). *The Fractal Geometry of Nature.* New York: W. H. Freeman & Co.

Maor, E. (1994). *e: The Story of a Number.* Princeton, N.J.: Princeton University Press.

Minnaert, M. (1993). *Light and Colour in the Outdoors.* New York: Springer-Verlag.

Melzak, Z. A. (1976). *Mathematical Ideas. Modeling and Applications.* New York: John Wiley.

Muir, J. (1996). *Of Men and Numbers: The Story of the Great Mathematicians.* New York: Dover Publications.

Mutchler, C. K., and C. L. Larson (1971). "Splash amounts from waterdrop impact on a smooth surface." *Water Resources Research* 7:195–200.

Neill, W. (1993). *By Nature's Design.* San Francisco: Chronicle Books.

Newman, J. R., ed. (1956). *The World of Mathematics.* New York: Simon and Schuster.

Nicolls, B. (1987). "A sense of proportion." *Report of the 10th International Congress of Vexillology,* 138–142. Winchester Mass.: Flag Research Center.

Nussenzveig, H. M. (1977). "The theory of the rainbow." *Scientific American* 236:116–127.

Oerlemans, J., and C. J. van der Veen (1984). *Ice Sheets and Climate.* Dordrecht, Netherlands: D. Reidel.

Ogilvy, C. S., and J. T. Anderson (1988). *Excursions in Number Theory.* New York: Dover Publications.

Peitgen, H. O., and P. H. Richter (1986). *The Beauty of Fractals*. New York: Springer-Verlag.

Perelman, Y. (1982). *Physics Can Be Fun*. Moscow: Mir Publishers.

Pollard, A. H. (1973). *Mathematical Models for the Growth of Human Populations*. Cambridge, U.K.: Cambridge University Press.

Raisz, E. (1962). *Principles of Cartography*. New York: McGraw-Hill.

Rand McNally (1978). *Grand Atlas of the World*. Chicago: Rand McNally.

Resnikoff, H. L., and R. O. Wells, Jr. (1984). *Mathematics in Civilization*. New York: Dover Publications.

Ribenboim, P. (1995). *The New Book of Prime Number Records*. New York: Springer-Verlag.

Richardson, E. G. (1952). *Dynamics of Real Fluids*. London: Pitman.

Rizika, J. W. (1950). "A philomathic study of rain." *American Scientist* 38:247–252.

Schele, L., and D. Freidel (1990). *A Forest of Kings*. New York: William Morrow.

Schwartz, L. W. (1988). "Recent developments in Hele–Shaw flow modeling." In *Numerical Simulation in Oil Recovery*, ed. M. F. Wheeler. New York: Springer-Verlag.

Scorer, R. S. (1978). *Environmental Aerodynamics*. Chichester, U.K.: Ellis Horwood.

Sloane, N. J. A. (1973). *A Handbook of Integer Sequences*. New York: Academic Press.

Smith, W. (1975a). *Flag Book of the United States*. New York: Morrow.

—— (1975b). *Flags Through the Ages and Across the World*. New York: McGraw-Hill.

Snyder, G. S. (1984). *Maps of the Heavens*. New York: Abbeville Press.

Snyder, J. P. (1987). *Map Projections—A Working Manual*. USGS Professional Paper 1395. Washington, D.C.: Government Printing Office.

—— (1993). *Flattening the Earth: Two Thousand Years of Map Projections*. Chicago: University of Chicago Press.

Snyder, J. P., and P. M. Voxland (1989). *An Album of Map Projections*. USGS Professional Paper 1453. Washington, D.C.: Government Printing Office.

Song Jian and Yu Jingyuan (1988). *Population System Control*. Berlin: Springer-Verlag.

Soroka, W. W. (1954). *Analog Methods in Computation and Simulation*. New York: McGraw-Hill.

Steinhaus, H. (1969). *Mathematical Snapshots*. New York: Oxford University Press.

Struik, D. J. (1967). *A Concise History of Mathematics*. New York: Dover Publications.

Sverdrup, H. U., M. W. Johnson, and R. H. Fleming (1942). *The Oceans: Their Physics, Chemistry and General Biology*. Englewood Cliffs, N.J.: Prentice-Hall.

The World Almanac and Book of Facts (1994), New York: Pharos Books, Scripps Howard Co.

Thompson, D. W. (1961). *On Form and Growth*. Abridged edition, ed. J. T. Bonner. London: Cambridge University Press.

Thompson, J. M. T., and H. B. Stewart (1986). *Nonlinear Dynamics and Chaos*. New York: John Wiley.

United Nations (1993). *World Population Prospects 1992*. Department of International Economic and Social Affairs, Population Studies No. 135. New York: United Nations.

Vajda, S. (1989). *Fibonacci & Lucas Numbers and the Golden Section: Theory and Applications*. Chichester, U.K.: Ellis Horwood.

von Foerster, H., P. A. Mora, and L. W. Amiot (1960). "Doomsday: Friday, 13 November, A.D. 2026." *Science* 132:1291–1295.

von Seggern, D. H. (1990). *CRC Handbook of Mathematical Curves and Surfaces*. Boca Raton, Fla.: CRC Press.

Walker, J. (1987). "Fluid interfaces, including fractal flows, can be studied in a Hele–Shaw cell." *Scientific American* 287:134–138.

Wells, D. (1986). *The Penguin Dictionary of Curious and Interesting Numbers*. Hammondsworth, U.K.: Penguin Books.

Westing, A. H. (1981). "A note on how many humans that have ever lived." *Bioscience* 7:523–524.

Wright, J. W., ed. (1992). *The Universal Almanac 1993*. Kansas City, Mo.: Andrews and McMeel.

Index

DATE